Environmental Earth Sciences

Series editor

James W. LaMoreaux, Tuscaloosa, USA

More information about this series at http://www.springer.com/series/8394

Kenji Kashiwaya

Geomorphology of Lake-Catchment Systems

A New Perspective from Limnogeomorphology

Springer

Kenji Kashiwaya
Institute of Nature and Environmental Technology
Kanazawa University
Kanazawa, Ishikawa
Japan

ISSN 2199-9155 ISSN 2199-9163 (electronic)
Environmental Earth Sciences
ISBN 978-981-10-5109-8 ISBN 978-981-10-5110-4 (eBook)
DOI 10.1007/978-981-10-5110-4

Library of Congress Control Number: 2017943251

Printed on acid-free paper

This Springer imprint is published by Springer Nature
The registered company is Springer Nature Singapore Pte Ltd.
The registered company address is: 152 Beach Road, #21-01/04 Gateway East, Singapore 189721, Singapore

Preface

Geomorphology is an earth science of landform changes and earth-surface processes. The changes and processes have been essentially controlled with tectonic activities and climatic activities. Recently, anthropogenic activities have also been important agents for those processes. One of the ultimate objects of this particular earth science is to predict the altitude (z) of a certain point (x, y) on the earth's surface at a certain time (t). However, it is nearly impossible to make that prediction exactly, even if all present sciences are used because the landform itself is an extremely complex system. It is of great significance, however, to grapple with the issue for the development and elaboration of earth science.

Geomorphology of lake-catchment systems (limnogeomorphology) also aims to contribute to this aspect (prediction) of geomorphology. Proper process-understanding and exact quantitative data are required for estimating future landforms precisely. However, both process-understanding and available data are limited in some phenomena and intervals. That is, only limited landform changes in limited observation intervals are to be quantitatively discussed for future estimation, although landforms have been developed in the geological and historical timescales as well as in the present observation period. Long-term development of landforms is also essential for geomorphology. Therefore, recent process-understanding and available quantitative data have to be extended to historical and geological timescales.

One of the most important procedures for this process is to observe landform phenomena instrumentally and historically–geologically. The lake-catchment system is convenient for observing earth-surface changes both instrumentally and historically–geologically. Present lake-catchment processes and changes can be instrumentally observed, and the changes including surrounding environments may have been recorded in the lacustrine sediments. Generally, the lacustrine records include not only present observational information but also pre-observational (past, historically and geologically) information, and the two kinds of information would be continuous without natural and artificial disturbance. If appropriate relationships between instrumental data and lacustrine (proxy) data are established for the present observation interval, proxy data for the past may be available for quasi-instrumental ones. This could fill a "missing link" between process geomorphology and historical geomorphology.

The first step of discussion of why it is necessary to uncover the "missing link" (connecting instrumental data with proxy data) is to find a way to establish long-term external forces for checking the model on the development of drainage density (Chap. 2). The forces are simply expressed as a function of long-term climatic changes. In a non-glaciated region such as Japan, the precipitation factor (discharge due to precipitation) must be the most promising means for the external force among several climatic factors. This is linked to research of Lake Biwa sediments (Chap. 3). In the 1980s when the model was under consideration, the oldest lacustrine sediments in the world were those obtained from Lake Biwa. The need for establishing appropriate functions for the climatic factors led to the Milankovitch theory (Chap. 9) and Lake Baikal and Lake Khuvsgul research (Chap. 3).

A clue to construction of the model for drainage density (Chap. 2) was derived from a rill network study in a slope system (Kashiwaya 1979; 1980). The model was established with many field observations and then was checked by model experiment in the laboratory before it was finally verified with field experiments (instrumental observations on an experimental slope in the field for 2 years). This suggests the importance of extension and connectivity of short-term laboratory experiments to comparatively long-term field experiments and leads to the idea that short-term instrumental observation should be linked to long-term field observation.

The first step for checking the idea was to make sure of the possibility that the data obtained from lacustrine sediments in instrumental observation intervals in small lake-catchment systems are connected to ones derived from pre-instrumental observation ones at Lake Yogo (Chap. 3). The second step was, unfortunately, related to the Kobe earthquake in 1995. Just after the earthquake, two sediment traps were set on the floor of a small lake to obtain short-term deposited materials for measuring high-resolution physical properties of sediments and to compare measured items with core-sampled sediments (comparatively long intervals; beyond observational limits) (Chaps. 3 and 4). Furthermore, it was possible to establish some models for erosion transportation and sedimentary processes in a lake-catchment system if enough data were given with the measurement (Chaps. 5–7).

Appropriate pre-observational data with exact dates are very important for properly interpreting and reconstructing past processes and changes. Especially, proxy data in the historical period connected to those in the observational period require more precise dates when they are used for quantitative discussion with observational data. Some abrupt events with documentary records (e.g., large earthquakes, volcanic eruptions, nuclear bomb testing) may also be useful for establishing reasonable dates in addition to radiometric dating (e.g., ^{14}C, ^{10}Be, ^{210}Pb). Many seismic and volcanic events have occurred recently in Japan, and some of them, as well as the peak time of nuclear bomb testing, can be used for determining absolute age models in the historical period and the instrumental observation period for natural forces and anthropogenic forces on the earth's surface (Chaps. 3–5).

If a certain amount of appropriate data is obtained, quantitative expressions for changes in some earth-surface phenomena will be needed for postdiction and prediction. The first stage for quantitative expression is to establish empirical equations for the phenomena. The empirical equations might be partially developed into theoretical (causal) ones if the phenomena were limited to short time interval and a small space (Chaps. 6 and 7). However, most equations for the earth-surface phenomena will be improved with additional data and with causal relationships, keeping empirical expressions (Chaps. 6 and 8). Nevertheless, they are of great meaning because various significant factors among innumerable ones are screened for establishing more reasonable expressions with more cause-and-effect relationships.

The idea of publishing this book came to me when I was a visiting professor at the Department of Geography of National Taiwan University (NTU) in 2014. At that time there was a lecture titled "Limnogeomorphology," which was given to postgraduate and undergraduate students from NTU and NTNU (National Taiwan Normal University). Publications on "limnogeomorphology" or studies on lake-catchment systems from the geomorphological point of view were extremely limited then. Geomorphological studies on the systems, however, are of great significance, as explained above. Hence, I tried to compile research works mainly based on my own works up to that point and to add some new ideas or findings, first published in this book, for completing *Geomorphology of Lake-Catchment Systems*.

To my great sorrow, I lost four respected and intimate colleagues in the course of this study: Prof. T. Masuzawa of Nagoya University, Prof. K. Fukuyama of Mie University, Dr. D. Tomurhuu of the Institute of Geology and Mineral Resources, Mongolian Academy of Sciences (MAS), and Prof. T. Kawai of Nagoya University.

I am most grateful to Prof. JC Lin, his colleagues, and the Department of Geography of NTU and at NTNU. I am indebted to many colleagues for the publication of this book: Prof. Y. Tanaka of Kyung-Hee University (Korea), Prof. N. Hasebe and Dr. S. Ochiai of Kanazawa University, and Dr. JW LaMoreaux (Series Editor of Environmental Earth Sciences) for the critical reading of the first manuscript of this book; Dr. T. Itono of Kanazawa University for the checking of figures; Ms. U. Uyangaa of Kanazawa University (National University of Mongolia) for the interpretation of the Baikal map; and the many junior and senior colleagues who worked with me at Kanazawa University, Kobe University, and Kyoto University for new ideas and new data.

Research results included in this book were supported by many precursors: Prof. S. Okuda of Kyoto University, Prof. T. Okimura of Kobe University, and Prof. A. Yamamoto of Osaka Electro-Communication University. Overseas and domestic fieldwork was successfully completed with Chinese colleagues from the Chinese Academy of Sciences (CAS), Yunnan Institute of Geography, and Yanbian University; Russian colleagues from the Russian Academy of Sciences Siberia Branch (RAS SB); Mongolian colleagues from MAS and the National University of Mongolia; Taiwanese colleagues from NTU; Korean colleagues from the Korean Institute of Geology and Mineral Resources; and Japanese colleagues from the

Japanese Association for the Baikal International Research Program (JABIRP), Lake Biwa Research Institute, Osaka City University, Toyama University, Tateyama Caldera Sabo Museum, and Hokkaido University of Education.

I am also grateful to many close colleagues who established the Japanese Geomorphological Union and those who have joined the East Eurasia Workshops since 2004 from Korea, China, Mongolia, Taiwan, Russia, and other areas. The financial support for the research work from Kanazawa University, the Japan Society for the Promotion of Science (Grants-in-Aid for Scientific Research), and the Strategic International Research Cooperative Program of the Japan Science and Technology Agency (JST) are gratefully acknowledged.

Kanazawa, Japan
March 2017

Kenji Kashiwaya

Contents

Chapter 1
Introduction

Studies on earth-surface processes and landforms are closely related in geomorphology, which is one of important features, compared with other earth sciences. In other words, "differential" and "integral" studies are inseparably connected. These two disciplines may correspond to process geomorphology (dynamic geomorphology) and historical geomorphology, respectively. The relation between differential equations expressing material transport and integrated landforms describes the temporal changes in landforms (historical geomorphology) on the basis of their developmental processes (process geomorphology). This approach was first described as theoretical geomorphology in the early 1960s (e.g., Scheidegger 1961).

There are some areas in geomorphology which try to combine "present" science (based on instrumental measurements and/or laboratory experiments) with "historical" science (with qualitative field observations) (theoretical geomorphology, physical geomorphology, etc.) (e.g., Scheidegger 1961; Ahnert 1987; Mizutani 1999; Pelletier 2008, etc.). However, most areas lack past quantitative records, especially data suitable for quantitative discussion. One approach overcoming this situation relates to the geomorphology of lake-catchment systems, i.e., limnogeomorphology. Lacustrine sediments are recording media, and most phenomena happened in the systems are recorded with the media as well as measured with instruments. The time interval and precision of the recorded data depend not only on the length of the recording media (sediment thickness) and their resolution (sedimentation rate) but also on the (instrumental) observational interval. Comparatively short-term lake-catchment processes and their recording mechanisms could be clarified mainly through instrumental observations. Long and high-resolution records with present observations fundamentally provide some clues to make clear past processes and postdict past changes, and they will show some ideas to predict future changes, depending on the record length and data precision.

Sedimentary records of lake-catchment systems include significant environmental information in addition to limnological and geomorphological ones. It is not easy to obtain useful information for the lake-catchment studies directly because many appropriate filters have to be applied to retrieve the specific information of

K. Kashiwaya, *Geomorphology of Lake-Catchment Systems*,
Environmental Earth Sciences, DOI 10.1007/978-981-10-5110-4_1

interest. However, this difficulty is linked to the advantage of this discipline (geomorphology of lake-catchment systems), because both "process" and "historical" information are explicitly included in the systems. Geomorphology should be one of integrated earth sciences where process geomorphology and historical geomorphology are essentially combined.

The term "limnogeomorphology," as roughly defined above, was first used at "Limno-geomorphology Laboratory" at the Institute of Nature and Environmental Technology, Kanazawa University, Japan (Kashiwaya et al. 2004). The term seems to have appeared first in the title of the journal at 2008 (Kashiwaya 2008). It has not always been widely used although it was explained comparatively in detail in the book published in 2015 (Kashiwaya et al. 2015). This is partly because papers directly written with the concept of limnogeomorphology are limited, although many papers related to geomorphology of lake-catchment systems have been published to this date. On the contrary, "limnogeology" in related academic field has been widely used since the first half of the twentieth century (Einsele 1938). Present volume aims to fill this gap.

Predictions relating to earth-surface environments are mainly based on numerical calculations using data from instrumental observations. Most available data for such calculations are limited in the period after the Little Ice Age (LIA), although some precious observational data exist also for the earlier centuries. For example, precipitation data for Seoul (Korea) start at the period of King Sejong of Joseon Dynasty (1418–1450) (Chosen Sotokufu Kansokusho 1917), and continuous data are available after the late eighteenth century. However, to make comparatively long-term predictions (for the next some centuries), obtaining data for various past climatic regimes is necessary, not only instrumental observation data for the warm period after the LIA but also historical proxy data. They are of particular interest for understanding the processes involved and for future projections under various regimes (climato-environmental regimes).

Both observational data with instruments and proxy ones are to be "observed" in the same "observatory" for establishing continuous records for the "observatory." In this respect, lake-catchment systems offer the possibility to acquire continuous records of various regimes and to interpret earth-surface processes with respect to environmental changes. Instrumental observations in such systems provide information on the underlying processes as well as on the environmental changes in progress. Lacustrine sediments in the system will give information on present (observational) and long-term environmental changes (proxy data). If the proxy data are smoothly (logically) connected to the observational data, it would be possible to obtain long continuous datasets for the past climate-environmental changes. Thus, lake-catchment systems may be considered as proxy observatories.

Next, let us discus how to connect proxy data with observational ones on the basis of scientifically reasonable procedures. There are some basic steps in dealing with data on earth-surface processes and temporal changes (Kashiwaya 2012).

(1) Clarifying relations between instrumental observation of lake-catchment processes and sedimentary records, and establishing "instruments" expressing the relationships in the present period: This is deeply related to an understanding of the lake-catchment mechanical processes and the recording process. Data recorded in sediments are compared in detail with hydrogeomorphological observations and measurements in the system; events included in sedimentary records should be quantified. Discussions on event recording process are also important. In addition, considering the validity and limitation of sedimentary records (proxies) in the system is important for appropriate use of the data. These lead to the establishment of some "instruments" (transfer functions). For example, a proxy rain gauge can be established using the physical properties of lacustrine sediments (Shimada et al. 2002). The "measured" data would then be compared with instrumental observation data to ensure their validity and limitation.
(2) Observation of past environments with the above instruments (quantitative reconstruction) based on sedimentary information and documents to delineate the earth-surface processes in the historical period: The established instruments may be extended with some modifications to the historical period if the measured proxy data are supported by documentary records and dated proxy data.
(3) Estimation of long-term environmental changes recorded in sediments, mainly related to cosmic-solar and orbital fluctuations with extended instruments for the pre-historical period: The instruments measuring the sedimentary records in observational and historical periods can be regulated and/or modified using environmental information related to cosmic-solar and orbital fluctuations. Furthermore, understanding of elementary processes in the present system is of great use for interpreting the information and estimating it properly.

Significant hints for understanding the processes in lake-catchment systems may be given through present instrumental and non-instrumental observations. Instrumental observations in present small lake-catchment systems may also be relevant to understand the elementary processes in the past systems. Model experiments and numerical simulations are also of great use for investigating these processes and for establishing basic equations for elementary processes. Accomplishing the above steps is linked to establishing continuous quantitative data from the present to the past in the system. Nevertheless, this is difficult because of the complexity of the systems, especially in the long term. Some assumptions are also introduced at some stages in the steps, and they should be logically checked with data, if any, for establishing some approximate relations, which will be of great help to utilizing paleoenvironmental information for future predictions.

In this present text, some lake-catchment studies of various scales are introduced to develop the above-mentioned ideas; long-term information on environmental changes and processes in large lake-catchment systems are discussed and interpreted through this information with present instrumental observations and understandings in small systems. Finally, the basic concepts and methods of limnogeomorphology (geomorphology of lake-catchment systems) will be understood in this context.

References

Ahnert F (1987) Process-response models of denudation at different spatial scales. In: Ahnert F (ed) Geomorphological models, theoretical and empirical aspects. Catena Supplement, vol 10, pp 31–50

Chosen Sotokufu Kansokusho (Observatory, Government General of Korea) (ed) (1917) Chosen kodai kansoku kiroku chosa hokokusho (Research reports on paleo-observational records in Korea). Observatory, Government General of Korea, Incheon, 200p (in Japanese)

Einsele W (1938) Concerning chemical and colloid chemical reactions in iron phosphate systems for limnochemical and limnogeological viewpoints. Arch Hydrobiol 33:361–387

Kashiwaya K (2008) Studies on lake-catchment systems: an introduction to limnogeomorphology. Trans Japan Geomorphol Union 29:205

Kashiwaya K (2012) Earth surface processes and environmental changes in lake-catchment systems. Trans Japan Geomorphol Union 33:121–136

Kashiwaya K, Tsuya Y, Okimura T (2004) Earthquake-related geomorphic environment and pond sediment information. Earth Surf Proc Land 29:785–793

Kashiwaya K, Okimura T, Itono T, Ishikawa K, Kusumoto (2015) Present earth-surface processes and historical environmental changes inferred from lake-catchment systems. In: Kashiwaya K, Shen J, Kim JY (eds) Earth surface processes and environmental changes in East Asia–records from lake-catchment systems. Springer, Tokyo, 321p, pp 1–24

Mizutani T (1999) Butsuri-chikeigaku gaisetsu (Outline of physical geomorphology). Taimeido, Tokyo, 206p (in Japanese)

Pelletier JD (2008) Quantitative modeling of earth surface processes. Cambridge University Press, Cambridge, 304p

Scheidegger RE (1961) Theoretical geomorphology, Springer, Heidelberg, 333p

Shimada T, Kashiwaya K, Masuzawa T, Hyodo M (2002) Hydro-environmental fluctuation in a lake–catchment system during the late Holocene inferred from Lake Yogo sediments. Trans Japan Geomorphol Union 23:415–431 (in Japanese with English abstract)

Chapter 2
Lake-Catchment System and Drainage System

The drainage system is a dominant structure of catchments. Studies on drainage systems started in the 1940s and were related to the development of quantitative geomorphology in the USA. Horton (1945) was a pioneer in this area. He and his successors developed ordering systems for the drainage structure (Scheidegger 1965; Fig. 2.1), which was linked to the development of the theory of fractal (Mandelbrot 1977).

Considering the temporal and spatial development of a drainage system, and the continuous records of various climatic regimes at various scales, the following are discussed; (1) the fractal structure of the drainage system in a catchment, and (2) the compatibility of drainage systems in various zones ("ergodic" reasoning).

The fractal structure of the drainage system in a catchment is more understandable in the self-similarity between small-scale systems and large-scale ones; "cyclicity" in scales is of great importance for structural development in the system. A typical structure of drainage system is shown in Figs. 2.2 and 2.3 (Tokunaga 1978; Kashiwaya 1986). The cyclicity was first identified and described by Tokunaga (1978) (introduced by Tarboton 1996). The number of streams of each order ω within a basin of order Ω (Fig. 2.2) (generalization of Horton law) is given by;

$$N(\Omega, \omega) = \frac{Q^{\Omega-\omega-1} - P^{\Omega-\omega-1}}{Q-P}(2+\varepsilon_1 - P)Q + (2+\varepsilon_1)P^{\Omega-\omega-1}.$$

P and Q are, respectively, given by

$$P = \frac{2+\varepsilon_1 + \mathrm{K} - \sqrt{(2+\varepsilon_1+\mathrm{K})^2 - 8\mathrm{K}}}{2}$$

K. Kashiwaya, *Geomorphology of Lake-Catchment Systems*,
Environmental Earth Sciences, DOI 10.1007/978-981-10-5110-4_2

and

$$Q = \frac{2 + \varepsilon_1 + \mathrm{K} + \sqrt{(2 + \varepsilon_1 + \mathrm{K})^2 - 8\mathrm{K}}}{2},$$

with $\varepsilon_1 = {}_{\kappa}\varepsilon_{\kappa-1}$ and $\mathrm{K} = ({}_{\kappa}\varepsilon_{\lambda}/{}_{\kappa}\varepsilon_{\kappa-1})^{1/(\kappa-\lambda-1)}$, where ${}_{\kappa}\varepsilon_{\lambda}$ is the number of stream order λ entering a stream of order κ from the sides.

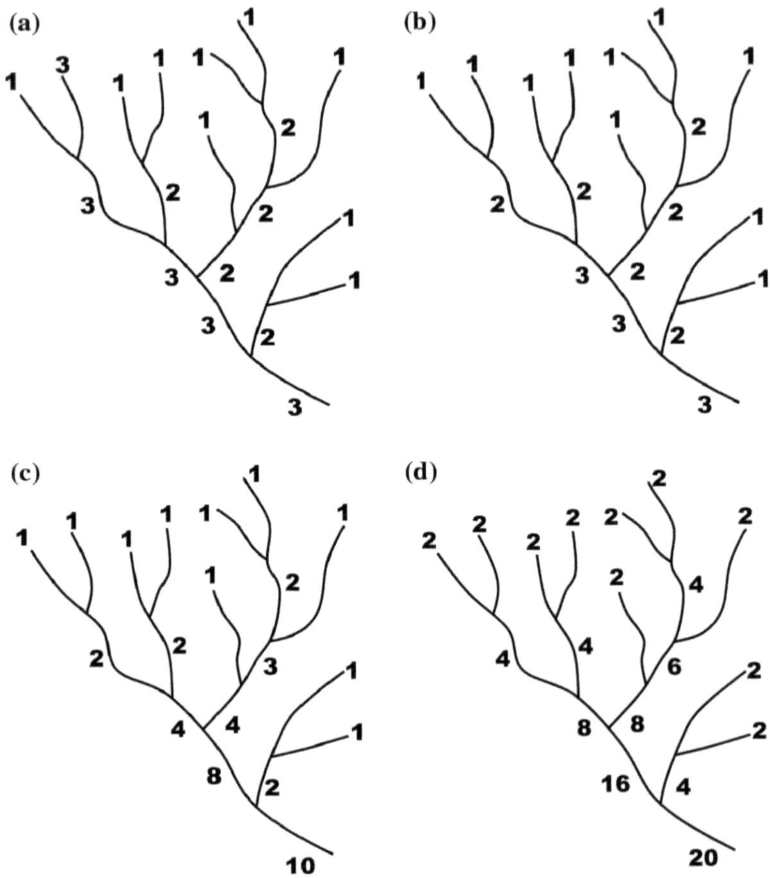

Fig. 2.1 Ordering method for streams proposed by **a** Horton (1945), **b** Strahler (1957), **c** Shreve (1966), and **d** Scheidegger (1965)

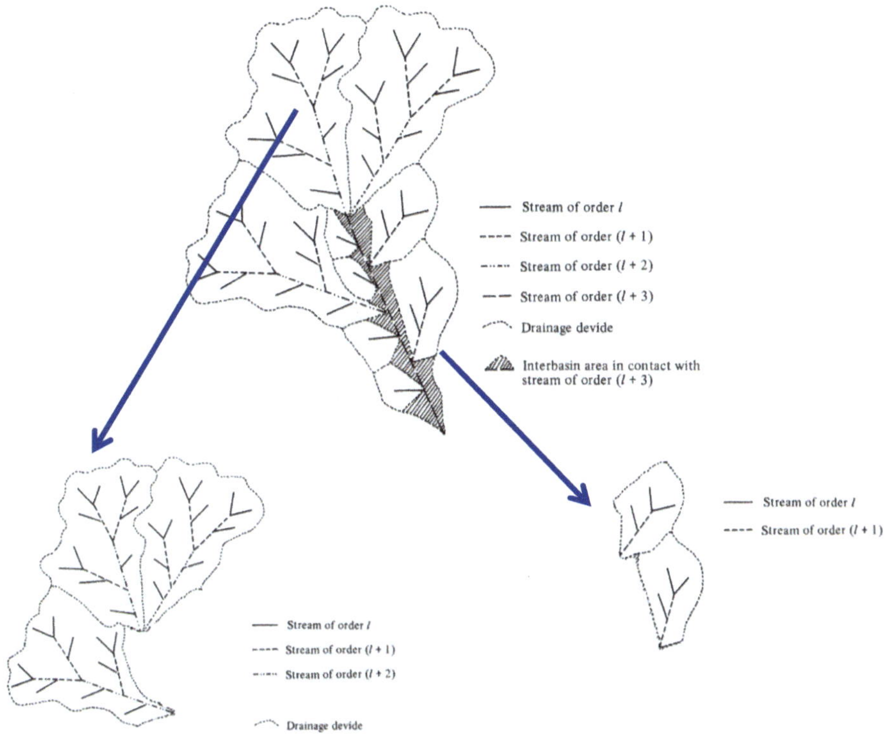

Fig. 2.2 Fractal structure of a generalized stream network (modified after Tokunaga 1978)

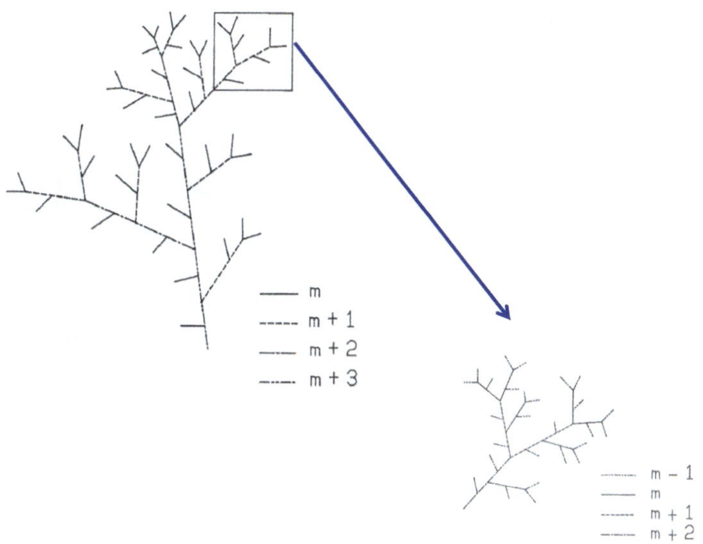

Fig. 2.3 Fractal structure of a small stream network (modified after Kashiwaya 1986)

Drainage systems in various stages under similar environmental conditions and structures are available for arranging a time sequence of a drainage system development; this is the "ergodic" reasoning in geomorphology (Paine 1985; Yatsu 1986). A mathematical model of the drainage system's development has been previously introduced and validated with field data (Kashiwaya 1983, 1987). A basic idea of this model was given from a rill network study in a slope system (Kashiwaya 1979, 1980). As is shown in Fig. 2.4, temporal change in drainage density is expressed by

$$\frac{\mathrm{d}D(t)}{\mathrm{d}t} = \beta'(t)\{1 - D(t)/\delta(t)\}D(t), \tag{2.1}$$

where $D(t)$ is the drainage density, $\beta'(t)$ is the relative erosional force, and $\delta(t)$ is the maximum drainage density determined by the physical properties of the catchment. It is solved, setting as an initial condition $D(t) = D_0$ for $t = t_0$

$$D(t) = \mathrm{e}^{-\int_{t_0}^{t}\beta'(\tau)\mathrm{d}\tau} \Bigg/ \left[\left\{\int_{t_0}^{t}(\beta'(\tau)/\delta(\tau)\mathrm{e}^{-\int^{\tau}\beta'(\varepsilon)\mathrm{d}\varepsilon}\right\}\mathrm{d}\tau + 1/D_0\right\}\right]. \tag{2.2}$$

This equation provides the temporal change of the drainage density in a lake-catchment system. In general, we have no long-term field data of the drainage

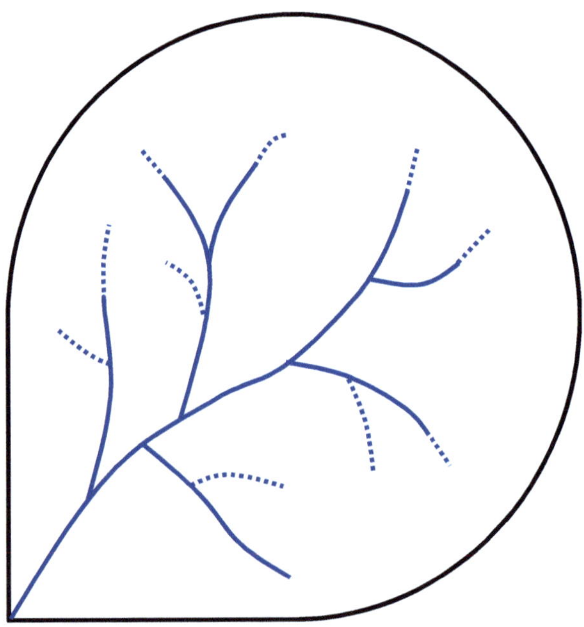

Fig. 2.4 An evolutional model of a stream network (modified after Kashiwaya 1986). Solid line; streams at time t. *Dotted line*; elongated segment of streams at time $t + \Delta t$

density, erosional force, and physical properties related to the system. Therefore, using dated catchment systems (or geomorphic surfaces), we assume that the drainage density in the dated systems or surfaces corresponds to the development stage of a certain catchment system, which is a typical example of the "ergodic" reasoning applied to catchment systems. In a hypothetical system, it is assumed that the function of the relative erosional force acting on the i-th system is expressed as $\beta_i'(t)$ and the river formation begins at t_i. Then, the present drainage density is expressed as follows:

$$\begin{aligned} D_1(t_1,T) &= \mathrm{e}^{-\int_{t_1}^{T}\beta_1'(\tau)\mathrm{d}\tau} \Big/ \left[\left\{\int_{t_1}^{T}(\beta_1'(\tau)/\delta_1(\tau)\mathrm{e}^{-\int^{\tau}\beta_1'(\varepsilon)\mathrm{d}\varepsilon}\mathrm{d}\tau\right\}+1/D_{01}\right\}\right] \\ D_2(t_2,T) &= \mathrm{e}^{-\int_{t_2}^{T}\beta_2'(\tau)\mathrm{d}\tau} \Big/ \left[\left\{\int_{t_2}^{T}(\beta_2'(\tau)/\delta_2(\tau)\mathrm{e}^{-\int^{\tau}\beta_2'(\varepsilon)\mathrm{d}\varepsilon}\mathrm{d}\tau\right\}+1/D_{02}\right\}\right], \\ &\cdots \\ D_i(t_i,T) &= \mathrm{e}^{-\int_{t_i}^{T}\beta_i'(\tau)\mathrm{d}\tau} \Big/ \left[\left\{\int_{t_i}^{T}(\beta_i'(\tau)/\delta_i(\tau)\mathrm{e}^{-\int^{\tau}\beta_i'(\varepsilon)\mathrm{d}\varepsilon}\mathrm{d}\tau\right\}+1/D_{0i}\right\}\right] \\ &\cdots \end{aligned} \quad (2.3)$$

where $D_i(t_i, T)$, $\beta_i'(t)$, $\delta_i(t)$, and D_{0i} are the drainage density, erosional force, maximum drainage density, and initial drainage density of the i-th system (geomorphic surface), respectively. If the relative erosional forces act equally on all systems (nearly under the same environmental (climatic) conditions) and the physical properties of the systems related to the erosional forces are nearly the same, we obtain

$$\begin{aligned} \beta_1'(t) &= \beta_2'(t) = \cdots = \beta_i'(t) = \cdots = \beta'(t) \\ \delta_1(t) &= \delta_2(t) = \cdots = \delta_i(t) = \cdots = \delta(t) \\ D_{01}(t) &= D_{02}(t) = \cdots = D_{0i}(t) = \cdots = D_0. \end{aligned} \quad (2.4)$$

Then, the drainage density of a series of systems during $T - t$ can be expressed in the same form:

$$D(t,T) = \mathrm{e}^{-\int_{t}^{T}\beta_1'(\tau)\mathrm{d}\tau} \Big/ \left[\left\{\int_{t}^{T}(\beta'(\tau)/\delta(\tau)\mathrm{e}^{-\int^{\tau}\beta'(\varepsilon)\mathrm{d}\varepsilon}\right\}\mathrm{d}\tau+1/D_0\right\}\right]. \quad (2.5)$$

It is necessary to give the functions $\beta'(t)$ and $\delta(t)$, D_0 in the equation and field data to evaluate this equation. However, establishing the functions and obtaining dated field data are difficult. Thus, it is assumed, as a first approximation, that $\beta'(t)$ and $\delta(t)$ are constant for this system:

$$D(t,T) = \delta D_0 \Big/ \left\{ D_0 + (\delta - D_0)\left(e^{-\beta'(T-t)}\right) \right\}. \tag{2.6}$$

This equation can be evaluated by using dated field data. Two examples are introduced: the drainage densities measured in terrace surfaces formed at different ages in Kanto, Japan (Kashiwaya 1983; Fig. 2.5a), and the drainage densities in drift surfaces formed at different times in North America (Kashiwaya 1987; Fig. 2.5b; Ruhe 1952). These support that the model introduced here is plausible,

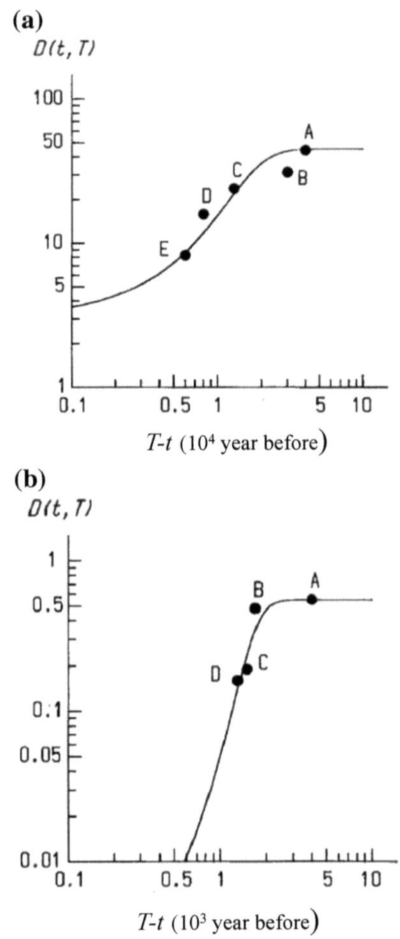

Fig. 2.5 **a** Temporal change in drainage density in the south Kanto, Japan. *Curve:* calculated results; *solid circles:* field data (*A:* 400 ka, *B*: 300 ka, *C*: 130 ka, *D*: 80 ka, and *E*: 60 ka); $D(t, T)$: drainage density (km/km^2); $T - t$: year. (modified after Kashiwaya 1983). **b** Temporal change in drainage density on glacial tills. *Curve:* calculated results; *solid circles:* field data (*A:* 40 ka, *B:* 17 ka, *C:* 15 ka, and *D:* 13 ka) given by Ruhe (1952). $D(t, T)$: drainage density (km/km^2); T–t: year. Reproduced from Kashiwaya (1987) by permission of John Wiley & Sons Ltd

although it is difficult to establish the functions and the physical properties related to D_{0i} (Fig. 2.4). However, knowledge of functions relating to external forces, weathering conditions, and the physical properties of materials constituting landforms ($\beta'(t)$, $\delta(t)$, D_0) is fundamental not only in limnogeomorphology, but also in geomorphology in general. Thus, sustainable efforts to obtain such knowledge are required.

References

Horton RE (1945) Erosional development of streams and their drainage basins: hydrophysical approach to quantitative morphology. Bull Geol Soc Am 56:275–370

Kashiwaya K (1979) On the stochastic model of rill development in slope system. Geogr Rev Jpn 52:53–65 (in Japanese with English abstract)

Kashiwaya K (1980) A study of rill development process based on field experiments. Geogr Rev Jpn 53:419–434 (in Japanese with English abstract)

Kashiwaya K (1983) A mathematical model for the temporal change of drainage density. Trans Jpn Geomorphol Union 4:25–31

Kashiwaya K (1986) Kasen (River) network. In: Shokokusha (ed) Katachi no himitsu (Mysteries of forms). Shokokusha, Tokyo, 248p, pp 99–116 (in Japanese)

Kashiwaya K (1987) Theoretical investigation of the time variation of drainage density. Earth Surf Proc Land 12:39–46

Mandelbrot BB (1977) Fractals: form, change and dimension. Freeman, San Francisco, p 365

Paine ATM (1985) 'Ergodic' reasoning in geomorphology: time for a review of the term? Prog Phys Geogr 9:1–15

Ruhe E (1952) Topographic discontinuities of the Des Moines lobe. Am J Sci 250:46–56

Scheidegger AE (1965) The algebra of stream-order numbers. US Geol Surv Prof Pap 525B:187–189

Shreve RL (1966) Statistical law of stream numbers. J Geol 74:17–37

Strahler AN (1957) Quantitative analysis of watershed geomorphology. Eos, Trans Am Geophys Union 38:913–920

Tarboton DG (1996) Fractal river networks, Horton's laws and Tokunaga cyclicity. J Hydrol 187:105–117

Tokunaga E (1978) Consideration on the composition of drainage networks and their evolution. Geogr Rep Tokyo Metrop Univ 13:1–27

Yatsu E (1986) Chikeigaku no ronri (Logic in geomorphology). In: Geographical Research Group of Chuo University (ed) Perspective of geography. Sozosha, Tokyo, 208p, pp 30–35 (in Japanese)

Chapter 3
Climatic (Climato-Geomorphic) Forces on Lake-Catchment Systems

Landform changes in a system are caused by natural (external and internal forces) and anthropogenic forces. There are three forces (climatic, tectonic, and anthropogenic impacts) acting on lake-catchment systems. First, climatic (climato-geomorphic) forces are examined.

3.1 Long-Term External Forces

Various temporal and spatial scales of these forces have acted and continue to act to deform earth-surface systems and landforms by flowing water, glacier, etc. Among the large-scale climatic changes, which are closely related to earth-surface environments, solar insolation is the first target for discussion. Climatic variability during the Quaternary has been explained mainly in terms of Milankovitch (Milanković) theory (1920, 1941). This theory was initially proposed by Milanković in 1920 but received little consideration in the 1950s and 1960s. The revival of the theory was initiated by Hays et al. (1976), who identified records of the Milankovitch cycles in oceanic sediments. These findings renew the interest in understanding the past long-term climatic (environmental) changes (see Chap. 9).

3.1.1 Lake Biwa System

The Milankovitch cycles were first identified in terrestrial (lacustrine) sediments of Lake Biwa, Japan (Yamamoto et al. 1984) (Chap. 9). These findings also drove researches on earth-surface processes and hydroenvironmental changes in lake-catchment systems because grain size was used as a proxy for precipitation although the age model for the core sediments was inadequate for statistical analyses in the earlier papers. However, clear Milankovitch cycles were identified afterward with revised age model for the core using a longer one (Kashiwaya et al.

K. Kashiwaya, *Geomorphology of Lake-Catchment Systems*,
Environmental Earth Sciences, DOI 10.1007/978-981-10-5110-4_3

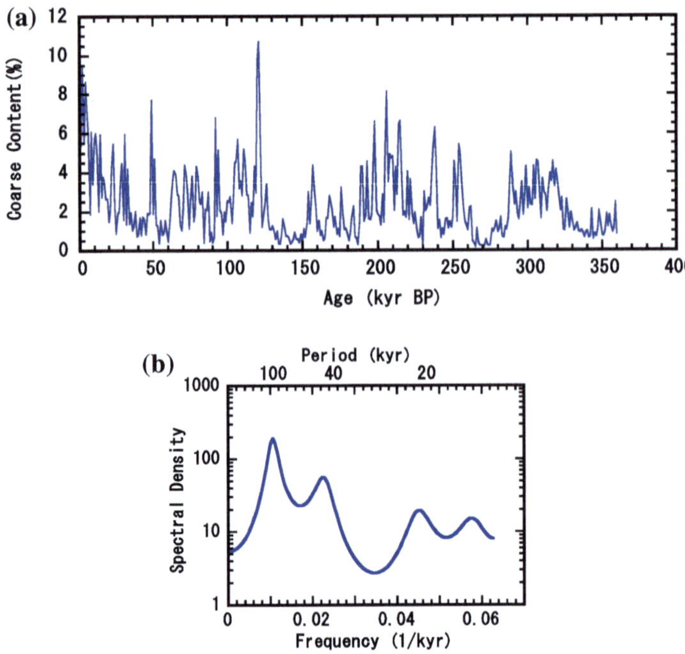

Fig. 3.1 **a** Temporal fluctuation in the coarse content of sediments from Lake Biwa, and **b** dominant periods in the fluctuation using spectral analysis

1991). Milankovitch cycles were clearly detected in the fluctuations of the coarse grain size content (more than 4.5 ϕ) in Lake Biwa sediments during the past 360 kyr (Fig. 3.1). These coarse grains include diatoms as well as minerals. Thus, the grain size content is a complex proxy for temperature, precipitation, and/or sediment discharge (catchment surface change) (see Chap. 9).

Furthermore, other physical properties (mineral grain size, mineral content, and bi-SiO_2 content) were discussed (Fig. 3.2). Temporal changes in mineral content and bi-SiO_2 content corresponded also to global climatic changes (increases in the mineral content corresponded to glacials, whereas increases in the bi-SiO_2 content were associated with interglacials and vice versa). The observed increasing trend in mineral grain size (median) is shown in the past 450 kyr (corresponded to the upper 250 m of the 1400-m core). This suggested a gradual increase in sediment discharge in the catchment and/or a decrease in the distance between the river mouth (inflow) and the sampling point (water level fall), as was experimentally indicated (Ochiai and Kashiwaya 2003). This was perhaps related to gradual aggradation and/or crustal movement in and around Lake Biwa (uplifting, distance shortening between river mouths and the sampling point). Long-term tectonic movements were also clearly detected in the long cores (200 and 1400 m) (Horie 1987). However, comparatively stable tectonic conditions (small events and/or gradual activity) were recorded in the upper silt–clay part (approximately 450 ka; 250 m), although some seismic activities were detected (see below).

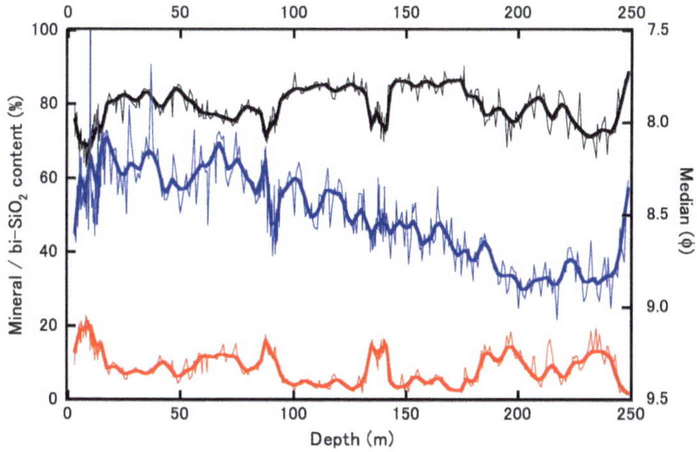

Fig. 3.2 Temporal changes in mineral content (*black line*), bi-SiO_2 content (*red line*), and mineral median (*blue line*) (original data by Ishikawa 2004)

To decipher the hydroclimatological fluctuation corresponded to the external force itself, it was necessary to detrend the observed trend included in the mineral grain size fluctuation. Figure 3.3 shows the changes in bi-SiO_2 content (temperature

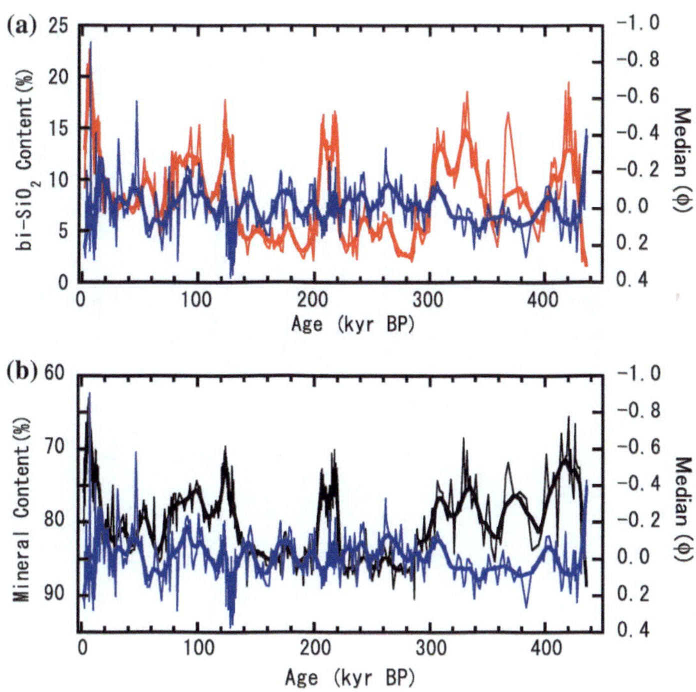

Fig. 3.3 **a** Temporal changes in bi-SiO_2 content (*red*), and detrended mineral median (*blue*; average value is assumed 0). **b** Temporal changes in mineral content (*black*), and detrended mineral median (*blue*). Bold ones; filtered

proxy) and the detrended mineral grain size (indicating precipitation, discharge, etc.) (Fig. 3.3a) as well as those in mineral content (indicators of aridity) and the mineral grain size (Fig. 3.3b). These indicate that the hydrological conditions were roughly modulated by temperature, but the difference between glacials and interglacials was not significant, perhaps because large discharge (large grain size) compensated the comparatively high water level (small grain size) during interglacials, whereas small discharge compensated the glacial low water level.

3.1.2 Lake Baikal System

Longer records obtained from Lake Baikal, Russia (Figs. 3.4, 3.5 and 3.6) included long climatic as well as tectonic information (Table 3.1). The climatic records indicated some differences between temperature (biological-activity) and discharge (precipitation) in the Pleistocene. Figure 3.7 shows the bi-SiO_2 content (temperature proxy) and mineral grain size fluctuations (precipitation/sediment discharge proxy) based on analyses conducted for the BDP98 core, which was obtained in the central ridge called the Academician Ridge (53°44′48″ N, 108°24′34″ E), in a comparatively shallow part of the lake (Ochiai and Kashiwaya 2005). Both bi-SiO_2 and mineral content fluctuations corresponded to global climatic change but yielded

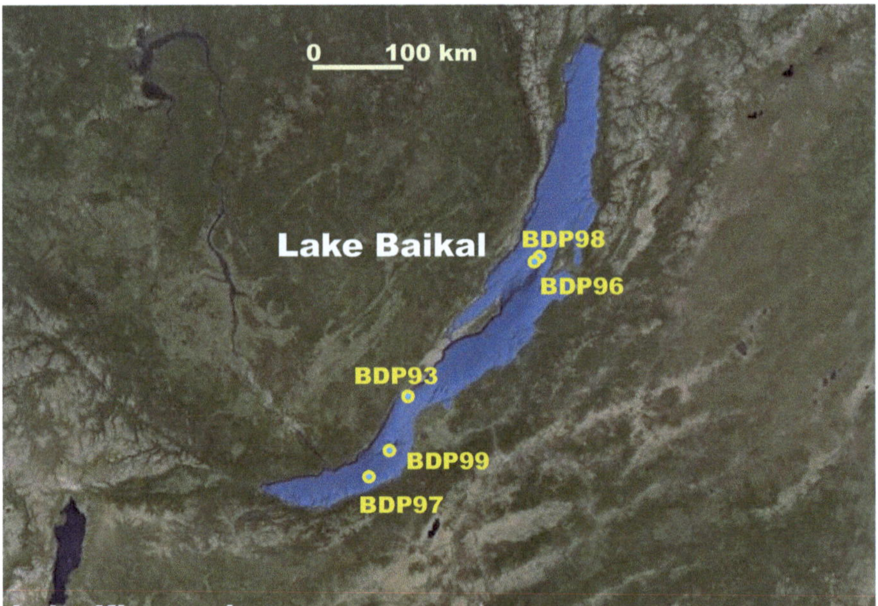

Fig. 3.4 Sampling points in Lake Baikal (Baikal Drilling Project) (modified after Google Earth)

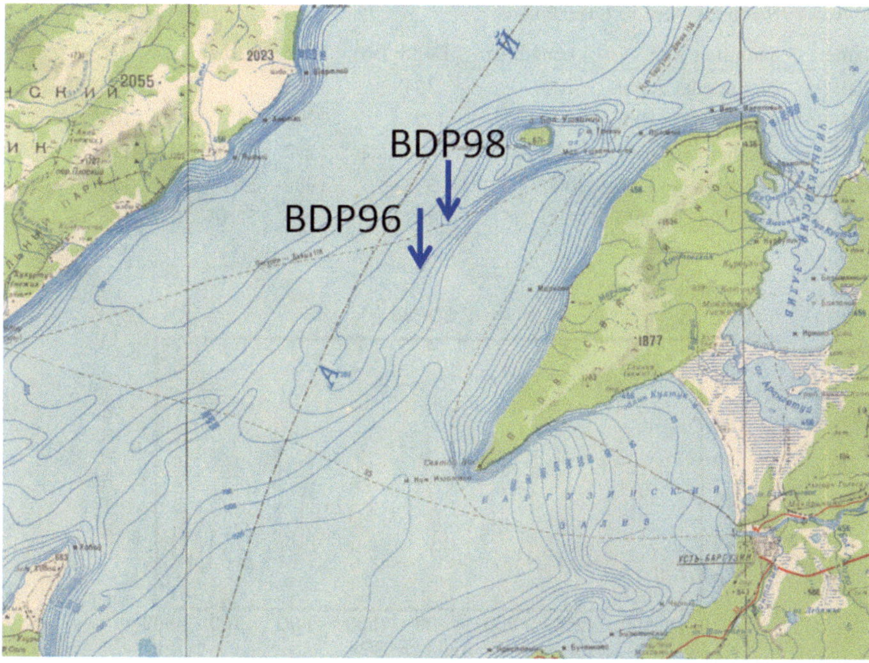

Fig. 3.5 Sampling points of BDP96 and BDP98. Modified after IKF (1996)

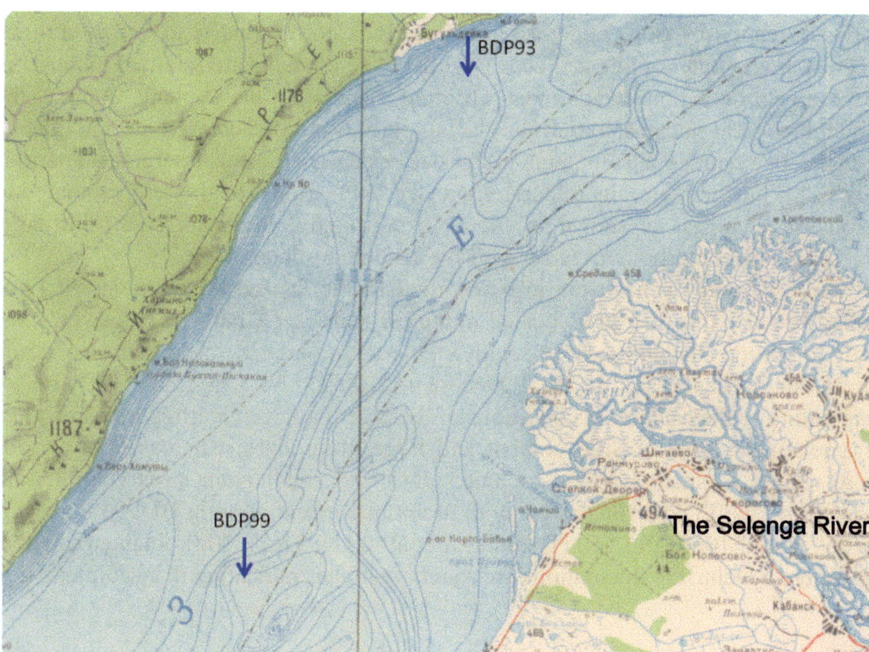

Fig. 3.6 Sampling points of BDP93 and BDP99. Modified after IKF (1996)

Table 3.1 Sampling sites in Lake Baikal

Name	Sampling site	Length (m)	Depth (m)	Year	Location
BDP-93	Buguldeika Saddle	100	354	1993	52°31′05″ N, 106°09′11″ E
BDP-96	Academician Ridge	200	321	1996	53°41′48″ N, 108°21′06″ E
BDP-97	Southern Basin	42	1436	1997	51°47′50″ N, 105°29′13″ E
BDP-98	Academician Ridge	600	333	1998	53°44′48″ N, 108°24′34″ E
BDP-99	Posolskaya Bank	300	201	1999	52°05′23″ N, 105°15′24″ E

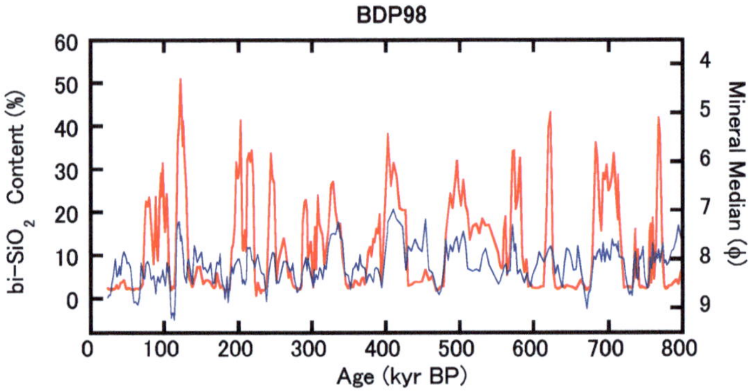

Fig. 3.7 Fluctuations in bi-SiO_2 content (*red*) and mineral median (*blue*) in the BDP98 core

opposite responses (low mineral content periods corresponded to warm periods, whereas high content indicated cold periods). The contents are relative; increase in the contents does not always mean absolute increase in the contents. Indeed, the mineral content increased during glacial periods and decreased in interglacial periods. In contract, the mineral grain size increased during interglacial periods (Fig. 3.8). Furthermore, the mineral grain size fluctuations roughly corresponded to insolation changes, as shown in Fig. 3.9; seemingly, the grain size (discharge, precipitation) could be correlated with the insolation better.

Many tectonic events were also recorded in the lacustrine sediments of Lake Baikal because this district is located in the rift zone (see Chap. 4). However, most high-amplitude fluctuations (especially, temperature-related ones) reflect external forces. In particular, in the BDP99 core (52°05′23″ N, 105°15′24″ E), directly responding to the catchment conditions of the Selenga River, warm periods resulted in a relatively large bi-SiO_2 content and large grain sizes, whereas cold periods were related to small bi-SiO_2 content values and small grain sizes (Fig. 3.10a). Comparing the PDP99 fluctuation with uranium concentration in BDP93 core (52° 31′05″ N and 106°09′11″ E) (Takamatsu et al. 2003; Fig. 3.10b) obtained offshore the Selenga Delta, the fluctuations are roughly similar, indicating that sedimentation

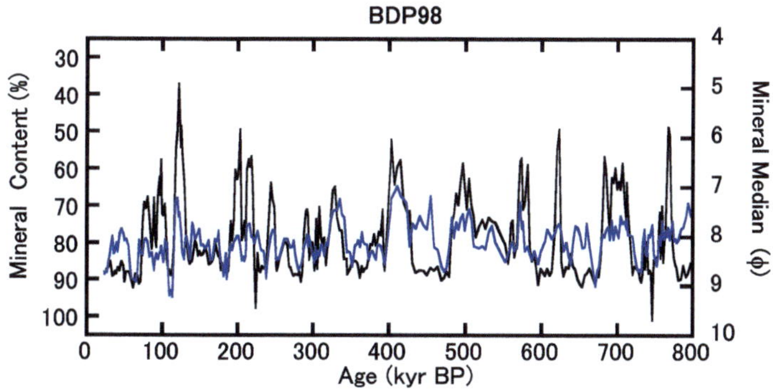

Fig. 3.8 Fluctuations in mineral content (*black*) and mineral median (*blue*) in the BDP98 core

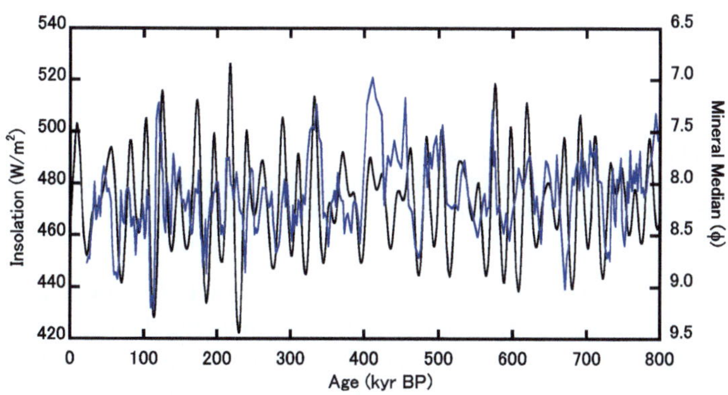

Fig. 3.9 Fluctuations in insolation (*black*) and mineral median (*blue*) in the BDP98 core

in both cases took place in similar environmental conditions. Consequently, it could be inferred that increase in the uranium concentration during warm periods corresponded to dissolution due to large water discharge (large precipitation and/or large melting water) in the uranium-rich area of the Selenga catchment (Edgington et al. 1996). In general, sediment production in the catchment resulted from terrestrial erosion and transportation, linked to landform change in the catchment. Hence, lacustrine sediments include information on the physical changes in the earth-surface environment of the catchment.

The lake-catchment system of Lake Khuvsgul (Hövsgöl) in Mongolia is another good example of the response to long-term climatic change (Kashiwaya et al. 2010; Fig. 3.4). One of the conspicuous characteristics of the system is the large water level difference (more than 100 m) between glacial and interglacial levels

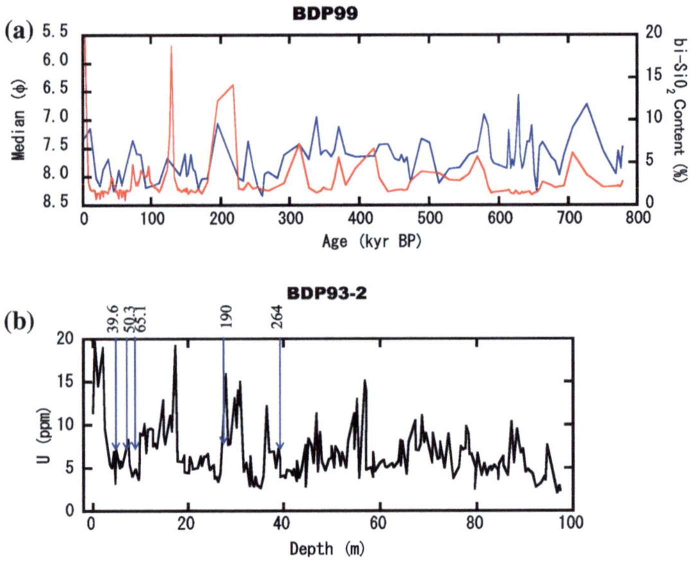

Fig. 3.10 **a** Fluctuations in median (*blue*) and bi-SiO_2 content (*red*) in the BDP99 core, and **b** fluctuation in uranium concentration (ppm) in the BDP93-2 core (modified after Takamatsu et al. 2003). Figures in the *upper side* of Fig. 3.10b indicate dates (ka)

(Krivonogov et al. 2005). HCl-soluble material (mainly $CaCO_3$) was identified as a good proxy for water level change, also corresponding to the global climatic changes of the past 240 kyr (Fig. 3.11). In addition, a large mineral content was related to cold (low water level) periods (except the mixed upper part). Generally,

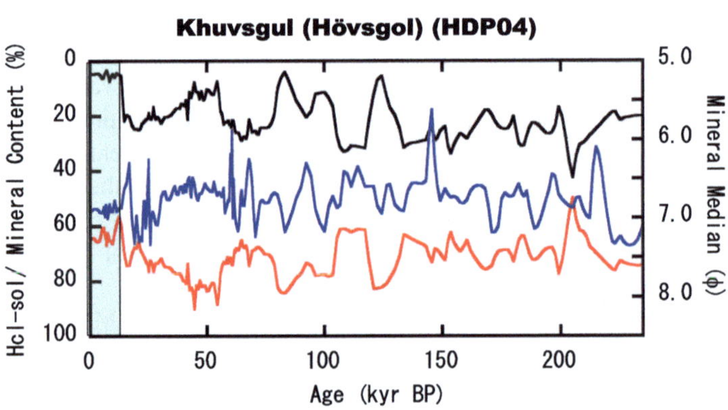

Fig. 3.11 Fluctuations in HCl-soluble ($CaCO_3$) content (proxy of water level; *black*), mineral content (*red*), and mineral median (proxy of discharge; *blue*) in the HDP04 core. Some mixed sediments may be included in the recent *light blue* part

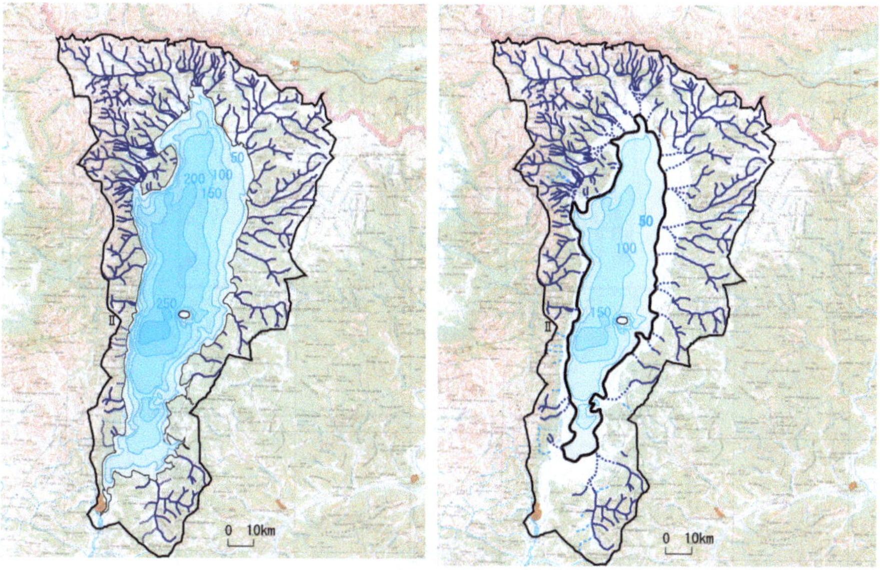

Fig. 3.12 Drainage density in the Holocene (*left*) and the late Pleistocene (*right*) (Yumoto 2010) in the catchment of Lake Khuvsgul

mineral grain size reflected sediment discharge in catchments under a stable water level. Apparently, grain size reflected not only sediment discharge but also water level change. Comparatively, large grain size was often detected in cold (dry) periods, possibly because river mouths approached the sampling point due to water level fall (Fig. 3.12; Yumoto 2010). This was also supported by the results (small grain size in the Holocene and large grain size in the late Pleistocene) obtained from a short core (X106) around HDP04 (Fig. 3.13; Tsukamoto 2005). As shown in the figures, this water level fall caused increases in the drainage density of the catchment and the sedimentation rate in the lake. The drainage system in catchments is of great importance for proper evaluation of the catchment evolution (e.g., Kashiwaya 1987), as discussed in the previous chapter. Table 3.2 (Yumoto 2010) shows the sedimentation rate for the Holocene and late Pleistocene epochs in the short-core samples obtained around HDP04 (Fig. 3.14), indicating high sedimentation rates during low water level periods (late Pleistocene) based on the mineral content.

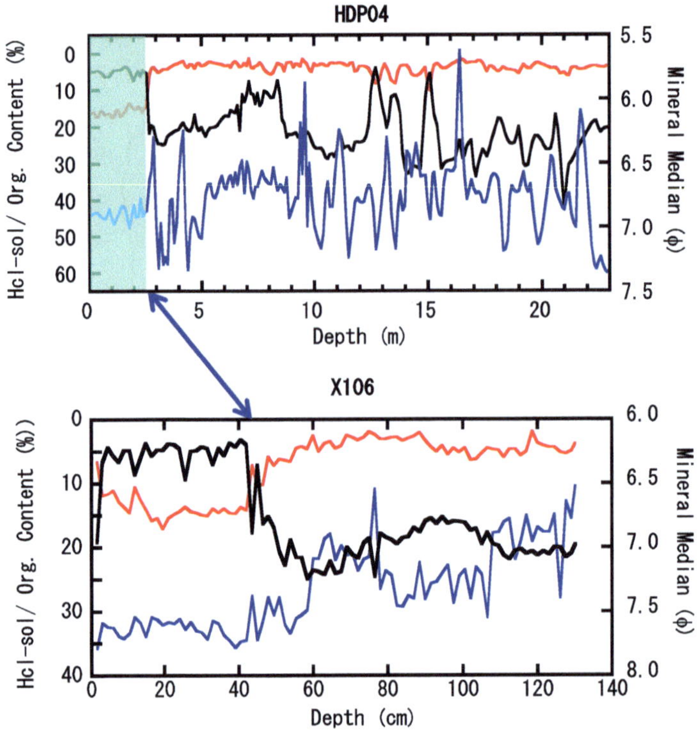

Fig. 3.13 Fluctuations in organic content (*red*), HCl-soluble ($CaCO_3$) content (*black*), and mineral median (*blue*) in HDP04 core (*upper*) and in X106 core (*lower*). The *upper light blue* part; mixed sediments (original data by Tsukamoto 2005)

Table 3.2 Sedimentation rate in the Holocene and late Pleistocene at several sampling points of Lake Khuvsgul (Tsukamoto 2005)

Core name	Sedimentation rate (cm/kyr) Holocene	Sedimentation rate (cm/kyr) Late Pleistocene
GC-1 J	0.2	7.1
GC-2 J	0.8	7.3
GC-4 J	1.2	6.3
X103	1.1	7.9
X104	1.0	8.0
X105	2.9	8.9
X106	2.3	6.0

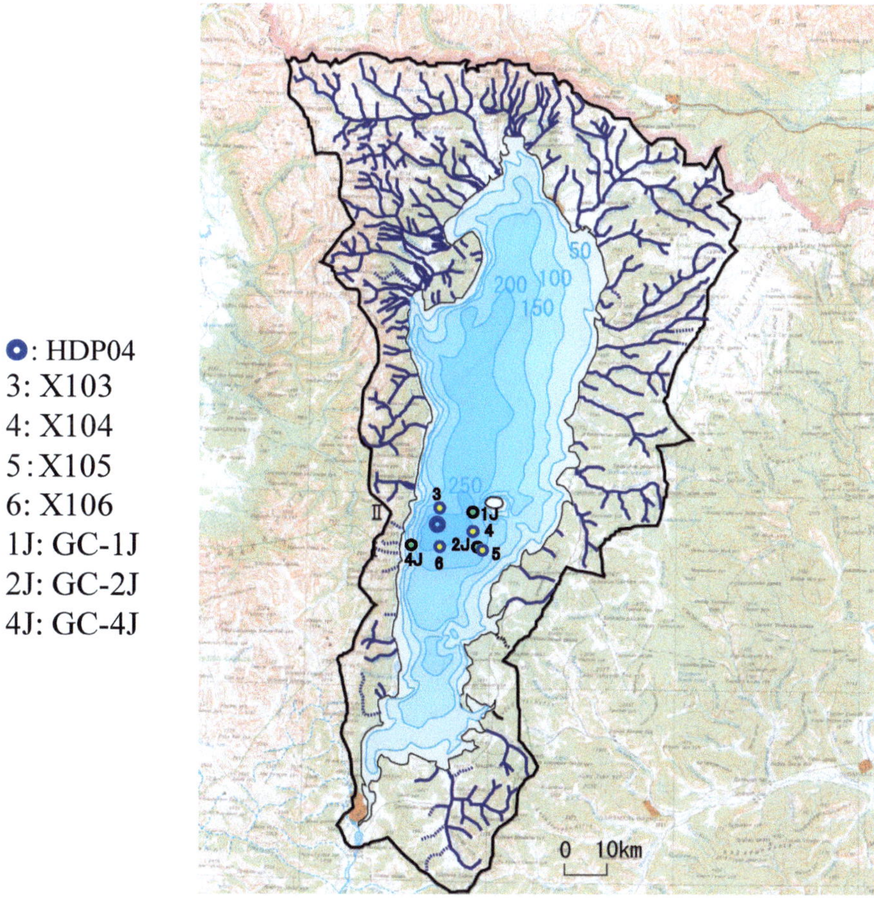

Fig. 3.14 Sampling points in Lake Khuvsgul

3.2 Short-Term External Forces

3.2.1 Lake Yogo and Lake Biwa

Information on lacustrine sediments in the present instrumental observational period can be used to clarify the relation between sediment records and meteorological settings and to understand the earth-surface processes. Here, two systems in central Japan are introduced as examples: Lake Yogo and Lake Biwa (Fig. 3.15). Lake Biwa is the largest lake in Japan. Lake Yogo is located north of the Lake Biwa catchment (Fig. 3.15). In Lake Yogo, several long cores (4 m cores) and short cores (1 m cores) have been obtained since the late 1990s because the lake environments are sensitive to changes in the surrounding catchment due to its compact system. An artificial flume was constructed around 1960, which enlarged the lake's catchment

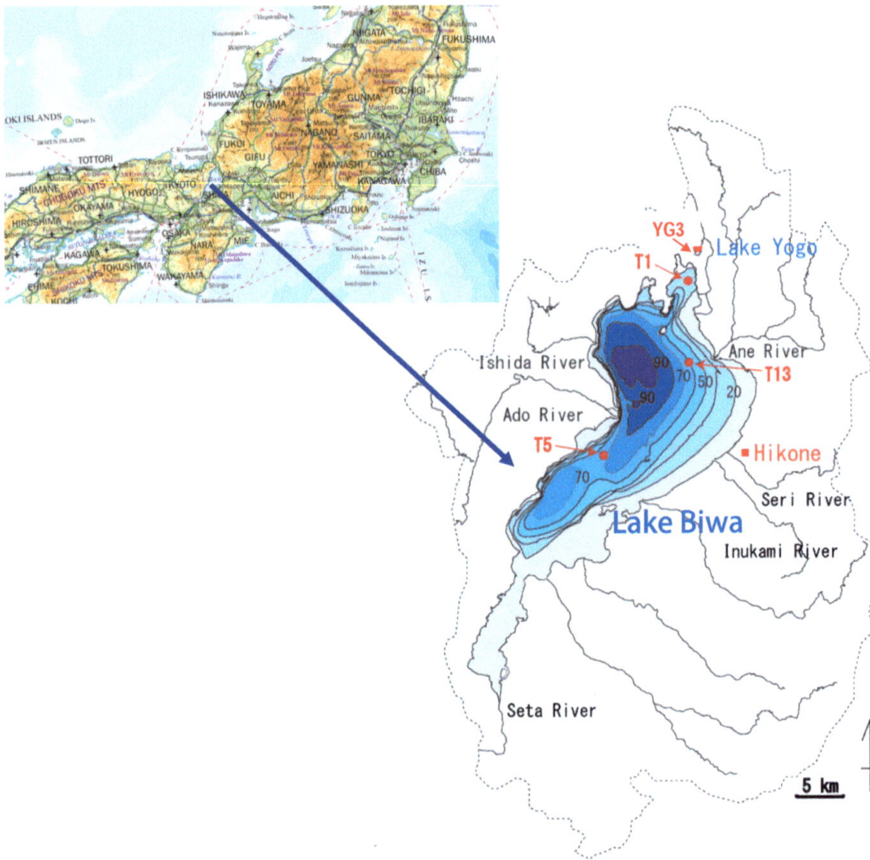

Fig. 3.15 Location of Lake Biwa and Lake Yogo (Maps from Geospatial Information Authority of Japan (GSI) and Itono et al. 2012)

area. The mineral grain size of a core obtained in the central part of the lake was highly proportional to the annual precipitation there (Fig. 3.16), indicating that the grain size could be used as a proxy for annual precipitation (proxy rain-gauge) (Shimada et al. 2002). Using this "gauge" (regression equation), the precipitation in the pre-instrumental observation period could be estimated, if an age model was properly established for the core. The age model proposed by the authors (Shimada et al. 2002) should be modified using the Kanbun earthquake event (1662), as discussed later, which is recorded in core samples (T-1) obtained in the northernmost part of Lake Biwa, close to Lake Yogo (Kashiwaya et al. 2015; Fig. 3.17). The peak in the mineral content (≈grain density) at the beginning of the large amplitude variation in grain size statistics (Figs. 3.17b and 3.18b) and the decreasing grain size (most recent large decreasing) was associated with the Kanbun earthquake in the T-1 core from Lake Biwa (green arrow in Fig. 3.17 and

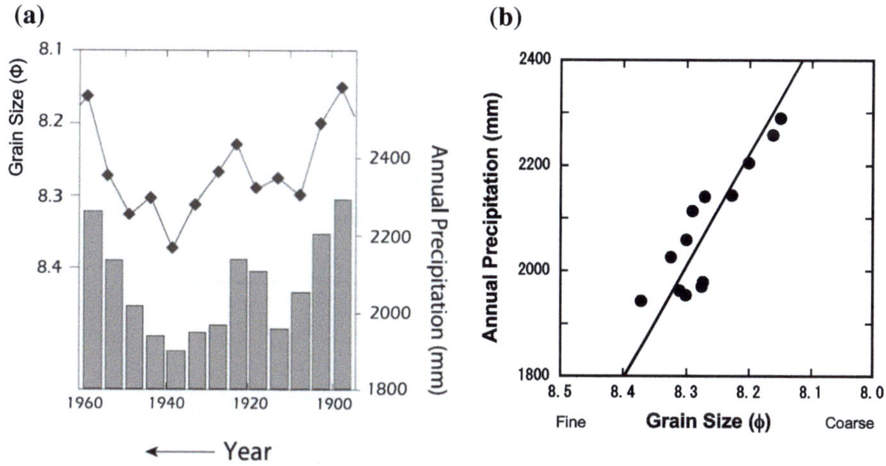

Fig. 3.16 **a** Changes in mineral grain size and annual precipitation, and **b** a relationship between mineral grain size and precipitation (Shimada et al. 2002)

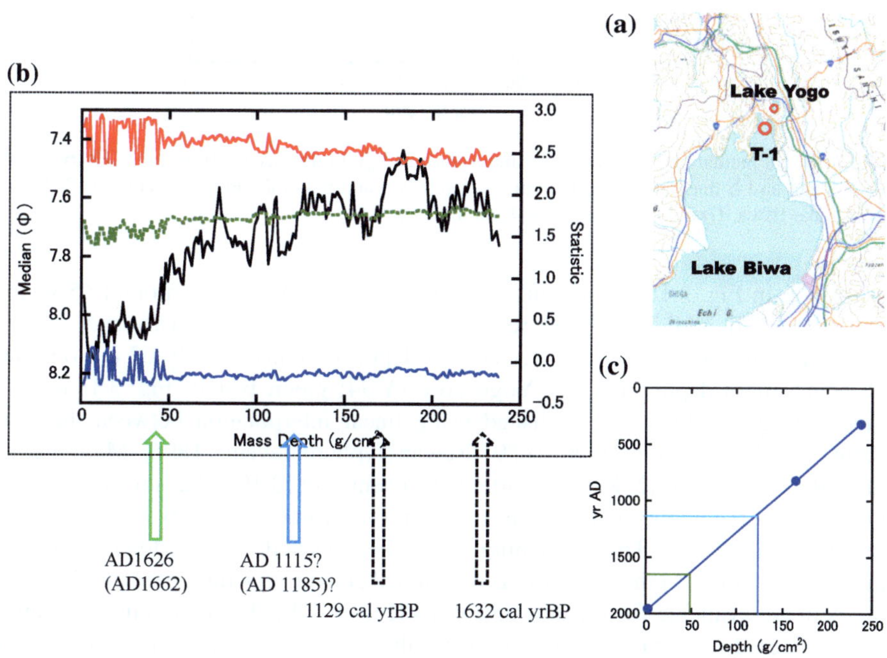

Fig. 3.17 **a** Location of Lake Biwa and Lake Yogo (Map from Geospatial Information Authority of Japan (GSI)), **b** changes in mineral grain size (*solid black*) and grain size statistic (*dotted green*; standard deviation, *solid blue*; skewness, and *solid red*; kurtosis) for the T-1 core, and **c** an age model for the T-1 core; two ^{14}C dates and one estimated from physical properties (modified after Kashiwaya et al. 2015)

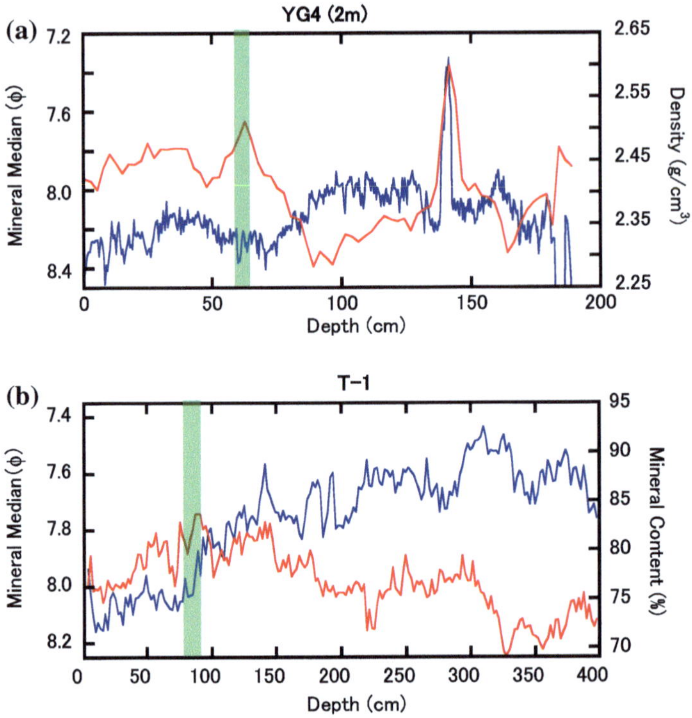

Fig. 3.18 **a** Fluctuations in mineral median (*blue*) and grain density (*red*) in the YG4 core from Lake Yogo and **b** fluctuations in mineral median (*blue*) and mineral content (*red*) in the T-1 core from Lake Biwa. *Green belts* mean the Kanbun earthquake

green belt in Fig. 3.18b at a depth of approximately 80 cm, i.e., around 45 g/cm^2 mass depth). In addition, a most recent large event (a peak in density and a lower depression in grain size) was detected at a depth of approximately 62 cm (around 18 g/cm^2 mass depth) in Lake Yogo core (YG4; green belt in Fig. 3.18a). The modified age model was established using linear interpolation between the four points: 1998, sampling time; 1959, Isewan typhoon event; 1896, Meiji heavy rainfall event; and 1662, Kanbun earthquake event (Fig. 3.19). The newly estimated precipitation including the recent historical (pre-instrumental observation) period is given in Fig. 3.20a. This estimation must be checked with different data for quantitative usage. In the Far East, as mentioned above, important precipitation data in Seoul are available for the past 250 years (Fig. 3.20b). The main climatic trends are considered similar in the Far East. Thus, the basic characteristics of precipitation in Seoul and central Japan (Hikone, the nearest JMA (Japan Meteorological Agency) station located in Lake Biwa coast) do not seem to significantly differ, and this is supported by the filter analysis (Fig. 3.21) and the spectral analyses (Fig. 3.22) for the two sites (Seoul and Hikone) in the interval of the instrumental observation period at Hikone. There is an almost linear relation between Hikone's

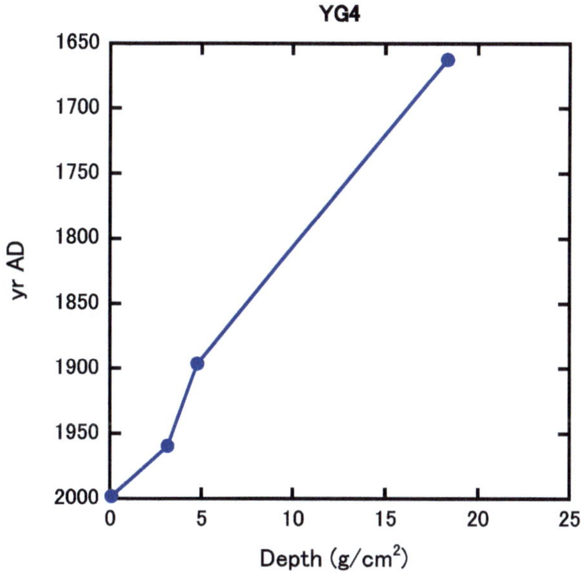

Fig. 3.19 Age model for Lake Yogo. *Solid circles* 1998, sampling time; 1959, the Isewan typhoon event; 1896, the Meiji heavy rainfall event, and 1662, the Kanbun Earthquake event

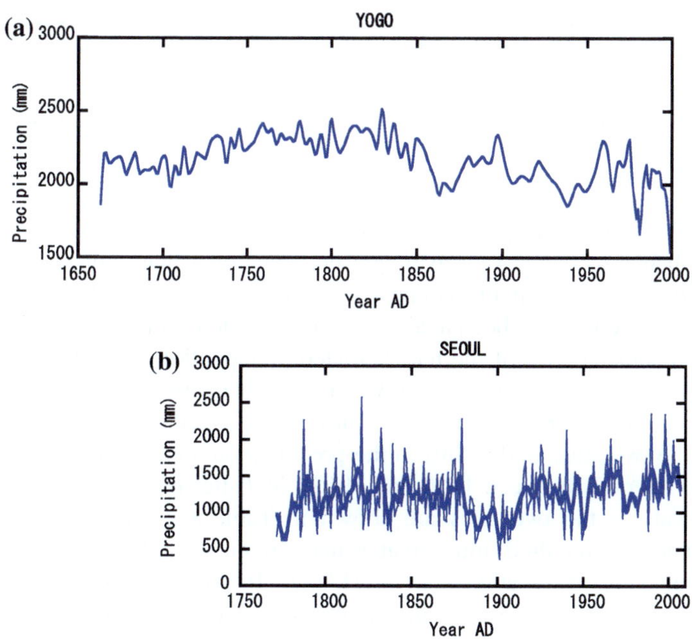

Fig. 3.20 **a** Estimated precipitation in Lake Yogo and **b** observed (*thin*) and filtered (*thick*) precipitation in Seoul

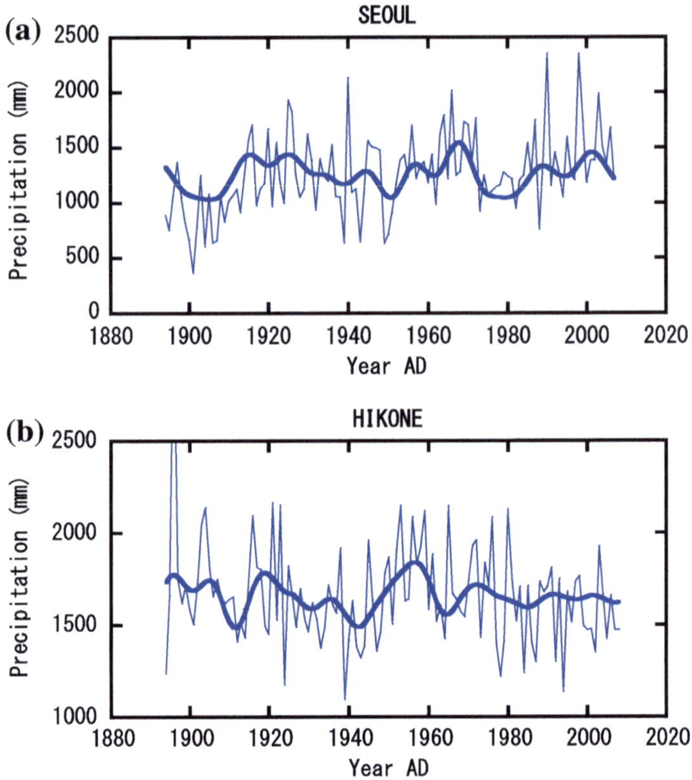

Fig. 3.21 **a** Observed (*thin*) and filtered (*thick*) precipitation in Seoul and **b** observed (*thin*) and filtered (*thick*) precipitation in Hikone

and Yogo's annual precipitation although the reconstructed Yogo's precipitation was established as a five-year-averaged precipitation value, considering the limitation of the sliced sediment thickness (Shimada et al. 2002). Quantitative precipitation data in Seoul have been available since the late LIA (since 1770). Therefore, the major characteristics of the reconstructed precipitation in Yogo since then can be compared with the Seoul data. However, Yogo's dataset is a filtered dataset. The annual precipitation in Seoul is shown also in Fig. 3.20b; there is a similar trend between the two sites, and a shift in the precipitation appears around 1880 at both sites (two intervals between 1750 and 1965). Statistics of the precipitation for the two intervals in the two sites are shown in Table 3.3. Yogo's precipitation is approximately a hundred mm greater than Seoul's, in both the pre- and the post-instrumental observation start at Hikone. In addition, during the late LIA, precipitation was a little larger than that after the LIA at both sites. In fact, standard deviation difference between the two sites is related to that between the observational data (Seoul) and the filtered estimated data (Yogo). For further discussion,

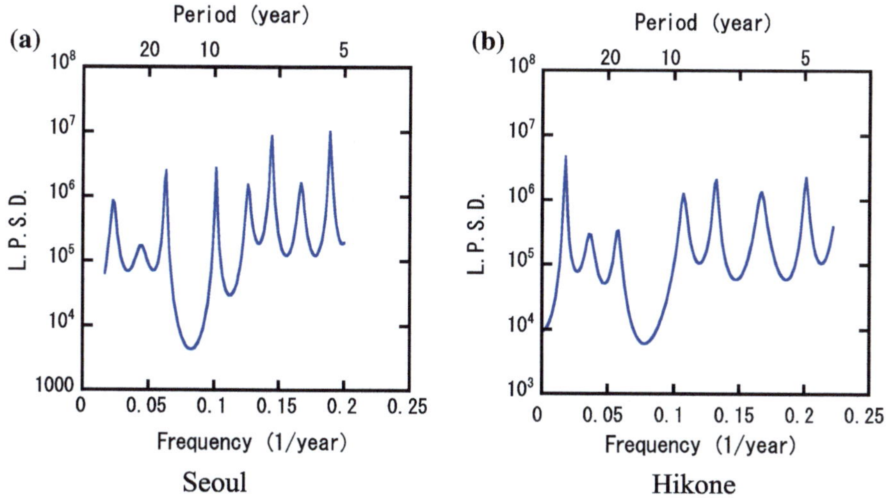

Fig. 3.22 Spectral analysis for **a** Seoul and **b** Hikone. Both calculated for the instrumental observation interval of Hikone Regional Observatory

Table 3.3 Statistics of precipitation (mm) in Seoul and Yogo

	Seoul (1771–1880)	Seoul (1880–1963)	Yogo (1771–1880)	Yogo (1880–1963)
Max	2582.0	2135.1	2518.7	2268.8
Min	622.0	370.0	1931.4	1905.2
Ave	1223.7	1127.2	2251.1	2077.3
S.D.	390.2	340.1	135.1	85.4
Skew	0.8585	0.3946	−0.6908	0.2726
Kurt	0.7034	0.0842	−0.2764	−0.4849

dominant periods are checked for the two intervals (post-1880 and pre-1880) in Seoul and Yogo (Figs. 3.23 and 3.24).

There is no large difference in the basic long dominant periods between the pre-1880 and post-1880 intervals in Seoul, despite the small difference in short periods, indicating similar basic trends in the precipitation in both intervals. No large difference is also observed for the dominant periods at Yogo in both intervals. These support that the estimated precipitation in the pre-instrumental observation interval at Yogo is plausible; the age model proposed here is possible and large shifts in density (mineral content) and mineral grain size are of great importance for establishing a good age model in this area.

In general, large lakes are not as sensitive as small ones to environmental changes in the surrounding catchments. Lake Biwa is the largest lake in Japan and pioneering researches on long lacustrine sediments were initiated here in 1971 as mentioned before (Horie 1984). Various lacustrine studies were also conducted in

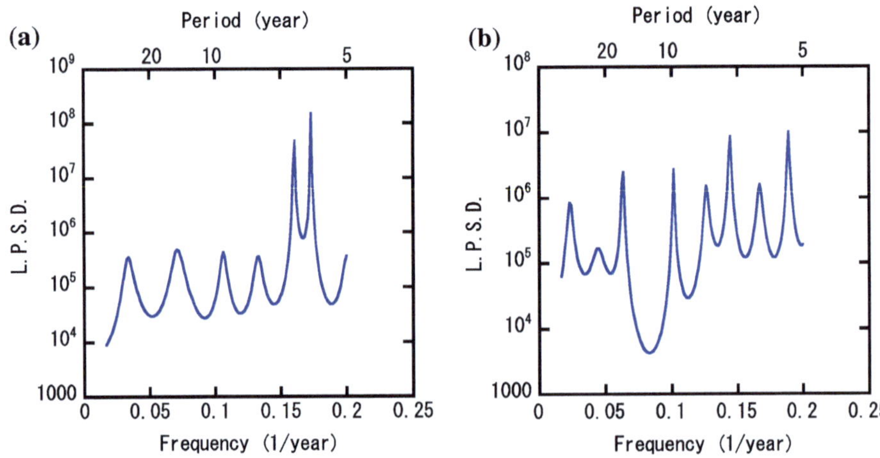

Fig. 3.23 Spectral analysis for the two intervals in Seoul **a** 1771–1880 and **b** 1880–1963

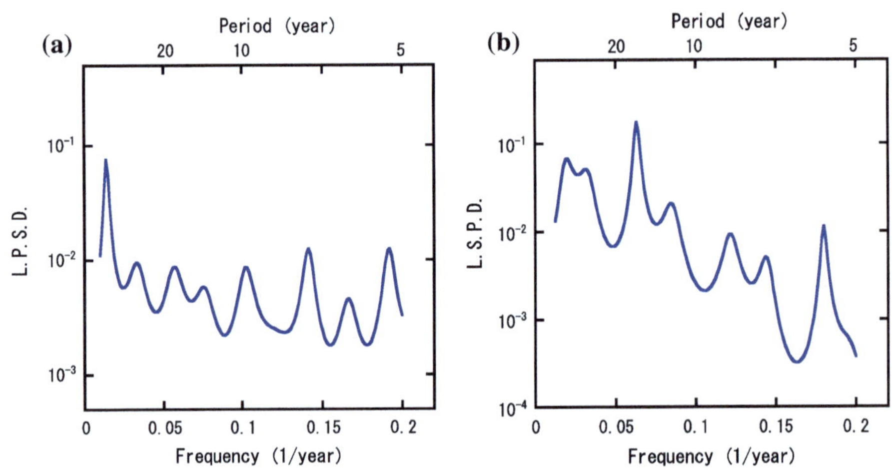

Fig. 3.24 Spectral analysis for the two intervals in Yogo **a** 1771–1880 and **b** 1881–1963

and around Lake Biwa. Researches on changes in recent environments (meteorological, climatological, hydrological, limnological, geomorphological, and vegetational conditions, etc.) with instrumental observation are of great use for modern limnogeomorphological studies (e.g., Okuda et al. 1995; Aota et al. 2006). Several short cores (1 m) were obtained at several sites in Lake Biwa at the end of the 1990s and the beginning of the 2000s aiming to investigate the lake-catchment processes and changes (Fig. 3.25). Moreover, some long cores (4 m) were examined by a joint team of researchers from Kanazawa University and Lake Biwa Research Institute. After Hikone Local Meteorological Observatory opened in

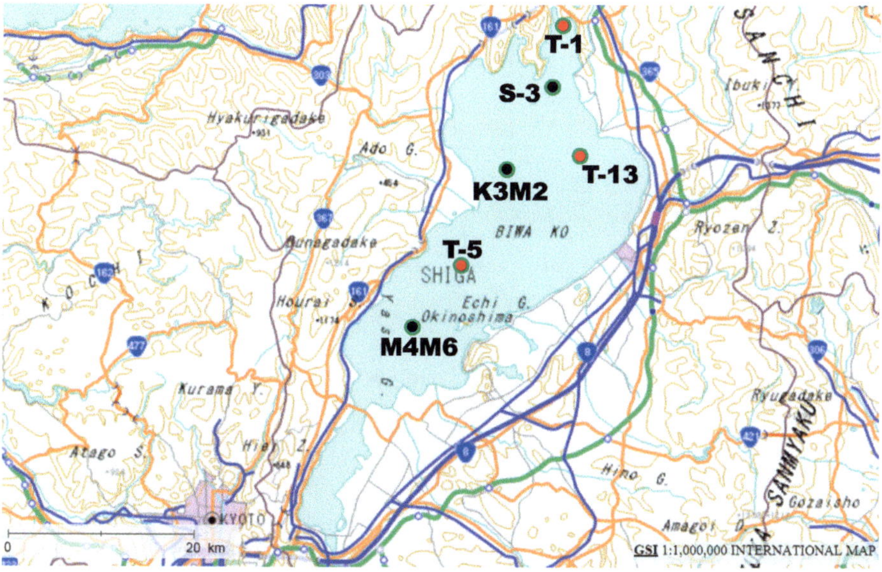

Fig. 3.25 Sampling points in Lake Biwa. *Circles* sampling points (based on GSI MAP; Geospatial Information Authority of Japan)

1893, Lake Biwa experienced two disastrous meteorological events: Meiji great heavy rainfall (1896) and Isewan typhoon (1959). The total rainfall for September 7–10, 1896, during the Meiji heavy rainfall was nearly 1000 mm (974 mm), and the maximum daily rainfall was 684 mm. The increase in the water level was 3.76 m, and all coastal zones were covered with water (Fig. 3.26).

The Isewan Typhoon, which passed along the south of Lake Biwa, attacked central Japan and killed more than 5000 people. Several floods happened also around Lake Biwa due to heavy rainfall in addition to the strong wind. It was the most catastrophic meteorological disaster in Japan in the instrumental observation period. Analytical results of the two physical properties (mineral content and density) for the core obtained at the northeast site (T-13) are shown in Fig. 3.27 (Itono et al. 2012). The two events can be detected in the physical properties (mineral content and density), using an age model using ^{137}Cs and ^{210}Pb dating and changes in 50-mm excess rainfall (50-mm excess rainfall refers to an annual summation of excess amount of 50-mm daily rainfall). Changes in the physical conditions of the lake-catchment due to heavy rainfall are expressed as sediment discharge, which is mainly related to mineral content, density, and mineral grain size. Figure 3.28 (Itono et al. 2012) shows changes in the three properties at the three sampling sites (T-1, T-5, and T-13 in Fig. 3.25). Two events are detected in the mineral content and the grain density in all cores (positive response).

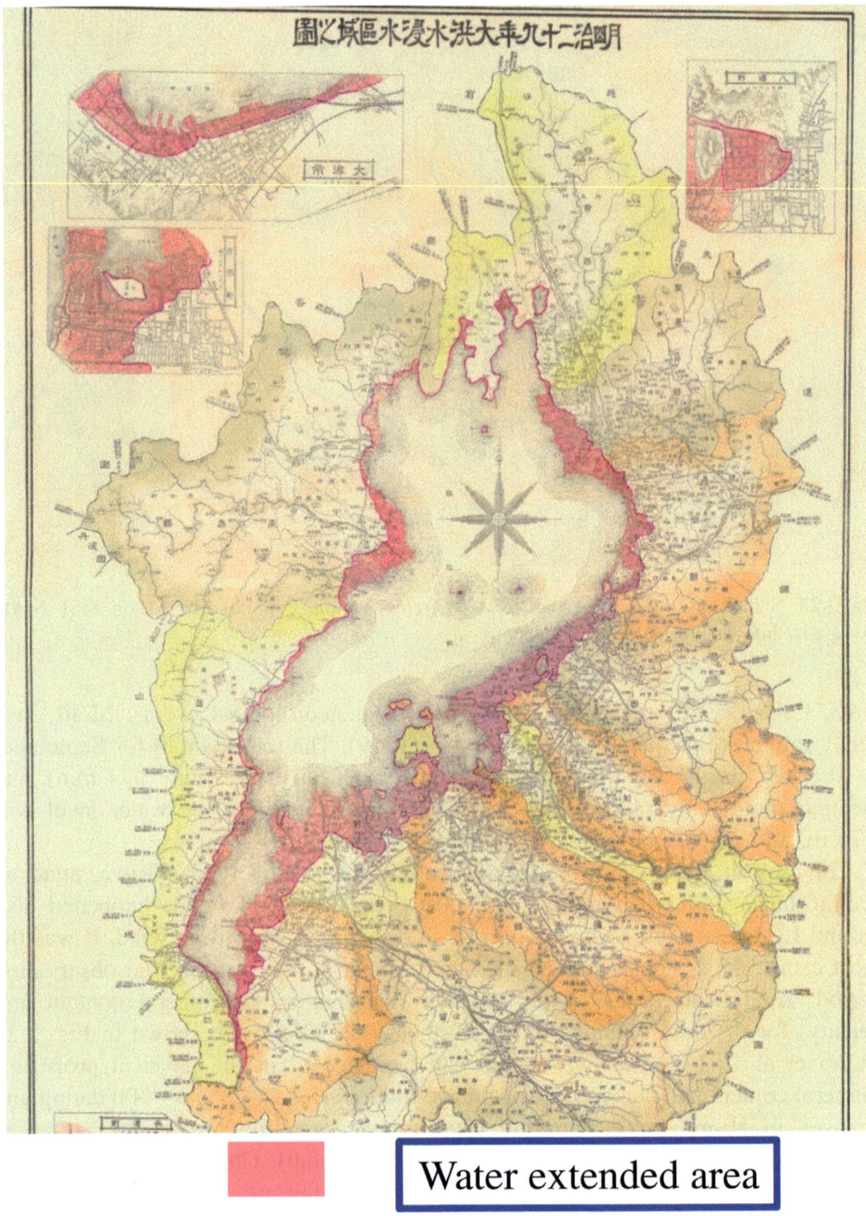

Fig. 3.26 High water level at the flood of 1896 in Lake Biwa system (modified after Biwako Chisui-Kai 1925)

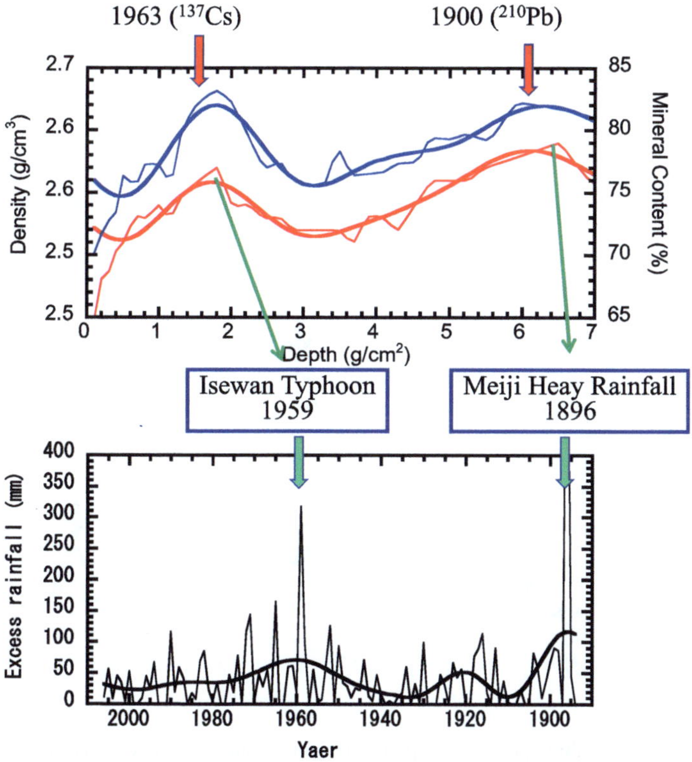

Fig. 3.27 Changes in physical properties (*blue*; mineral content, *red*; density) of sediments from the T-13 point (Fig. 3.25) in Lake Biwa (*upper*) and excess rainfall intensity (*black*) at the Hikone Regional Observatory (*lower*). *Red arrows* mean dates given by ^{137}Cs concentration and ^{210}Pb concentration. *Light green arrows* indicate disastrous events (the Meiji heavy rainfall; 1896, the Isewan Typhoon; 1959). *Bold curves* indicate filtered ones. Modified after Itono et al. (2012)

However, the Meiji heavy rainfall cannot be positively detected in the mineral grain size fluctuations in all cores (negative response), although the Isewan typhoon can be detected positively (T-13; fluvial-effect zone) and negatively (T-5; deeper flat zone). The number of landslides in the catchment due to the heavy rainfall was much larger during the Meiji heavy rainfall than during the Isewan typhoon (Shiga-ken 1966); however, it is impossible to compare the number of landslides between the two events because different survey methods were employed. A major comparable parameter is the water level change (coastal area change); the water level increase in the Isewan typhoon was 0.87 m. Generally, most coarse grains deposited in the river mouth and in the coastal zone. Transgression in the Meiji event was some kilometers in the inner catchment (Fig. 3.26). Most coarse grains

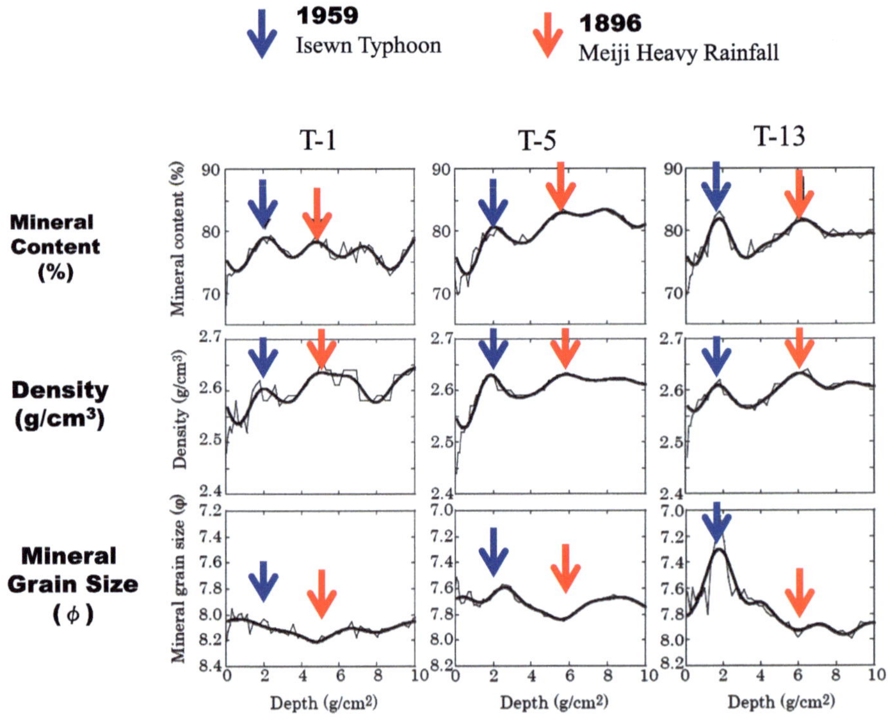

Fig. 3.28 Changes in three physical properties (mineral content, grain density, and mineral grain size) of sediments from the three points (T-1, T-5 and T-13 in Fig. 3.25) in Lake Biwa. *Blue arrows* mean the Isewan typhoon and *red arrows* mean the Meiji heavy rainfall. Modified after Itono et al. (2012)

may have deposited in the new river mouth and new coastal zone. To investigate this in detail, cores obtained in the deeper part were examined. Figures 3.29 and 3.30 show the mineral grain size fluctuation (blue) and the mineral content fluctuation (black) (Fig. 3.29) as well as the grain density (red) and the mineral content (black) (Fig. 3.30) in the deepest sites in the northern basin of the lake (K3M2 core) and in the central basin (M4M6 core). These indicate a slight decrease in the mineral grain size at the two events although the size was mostly finer than 8.3 ϕ, suggesting a slight different sedimentary condition from the coastal zone and fluvial-effect zone. The age model for the M4M6 core is based on the ^{137}Cs concentration shown in Fig. 3.31 and that for the K3M2 can be estimated from the same figure. The deepest sites generally constitute the final zone of deposition; comparatively fine materials, which were probably produced also in floods, may have settled in the zone, with some time lag for long traveling.

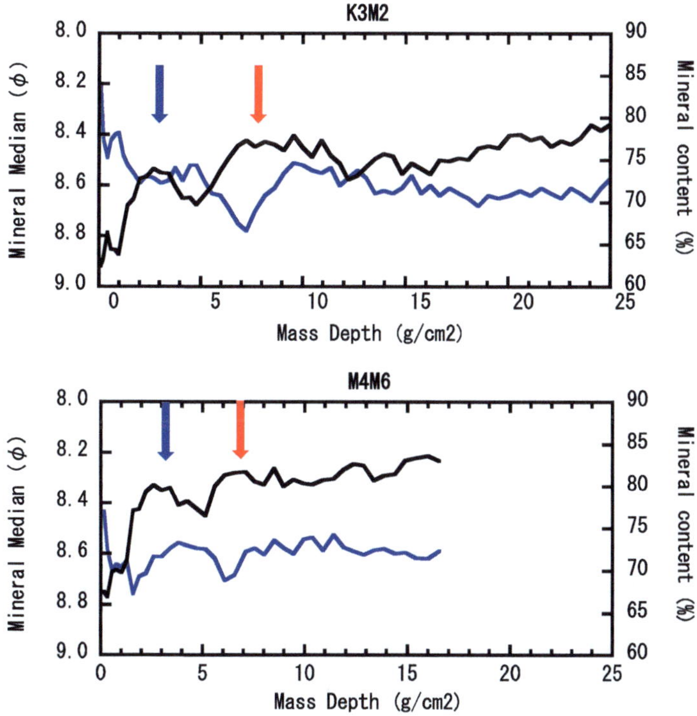

Fig. 3.29 Changes in physical properties (*black*; mineral content, *blue*; mineral grain size (ϕ) of sediments from the deepest part (K3M2 and M4M6 in Fig. 3.25) in Lake Biwa. *Red arrows* mean the Meiji heavy rainfall (1896) and *blue* ones mean the Isewan typhoon (1959). Original data by Ishikawa (2004)

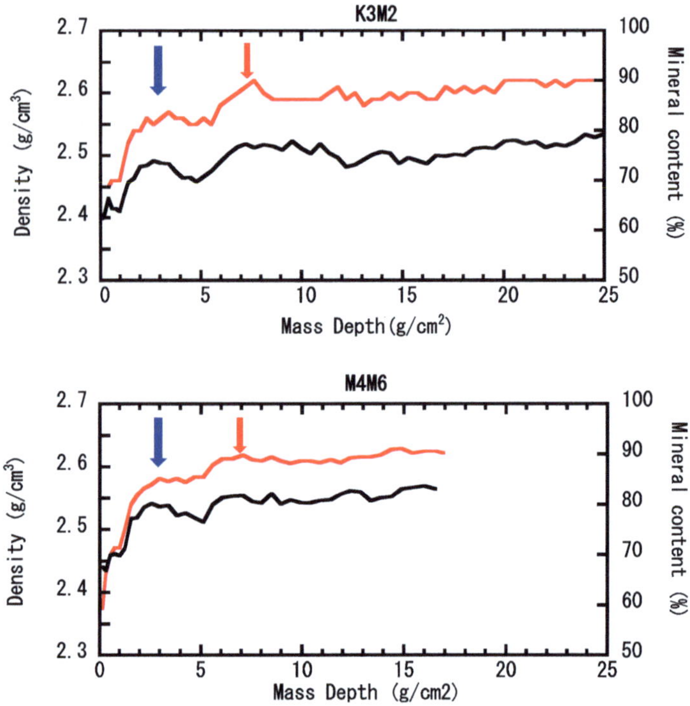

Fig. 3.30 Changes in physical properties (*black*; mineral content, *red*; grain density of sediments from the deepest part (K3M2 and M4M6 in Fig. 3.25) in Lake Biwa. *Red arrows* mean the Meiji heavy rainfall (1896) and *blue* ones mean the Isewan typhoon (1959) Original data by Ishikawa (2004)

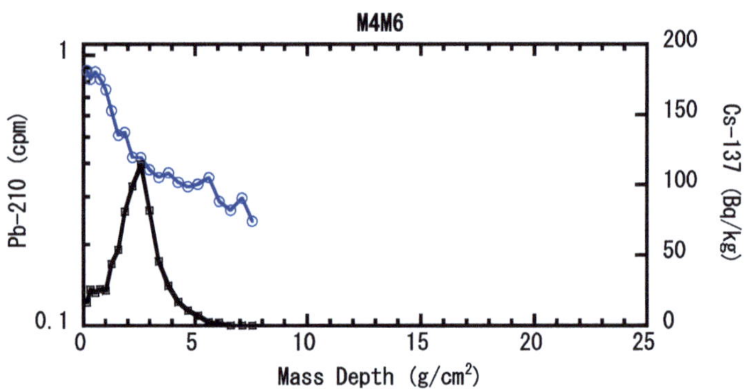

Fig. 3.31 Changes in ^{210}Pb concentration (*blue*) and ^{137}Cs concentration (*black*) for the M4M6 core. Original data by Ishikawa (2004)

3.2.2 *Kawauso-ike, Kobe*

Next, different geomorphological disasters due to heavy rainfall in a small lake (pond)-catchment system at Kobe (central Japan) are introduced (Fig. 3.32). Recently, Kobe is famous for its 1995 earthquake disaster (the relation between the earthquake and the lake-catchment system will be given afterward). Before the earthquake, the Kobe area was a major target of prevention research for landslide disaster due to heavy rainfall (e.g., Kashiwaya et al. 1986, 1988, 1995). Two disastrous heavy rainfall events occurred in 1938 and 1967 around Kobe in the instrumental observation period. During those events, many landslides and debris flows resulted from heavy rainfall (Figs. 3.32A–C, 3.33 and 3.34). Landslides in some areas occurred at nearly the same sites in the two events, suggesting landslide repetition due to weathering rate (Figs. 3.35 and 3.36, Kashiwaya et al. 1986, 1987). The Rokko Mountains consist of deeply weathered granite, and their surface materials are easily eroded. To investigate these events, several core sediments were obtained at Pond Kawauso-ike in the middle (575 m a.s.l.) of the Rokko Mountains, not so far from the active fault lines related to the 1995 Kobe Earthquake (E in Fig. 3.32).

The pond was said to have been used for irrigation of the paddy fields downstream. However, since the 1938 natural heavy rainfall disaster, this has not been

Fig. 3.32 Location of the sites studied (*A*, *B*, *C*, *D*, *E* and *F*) in Kobe, Japan. *Yellow dotted lines* mean active fault lines (modified after Google Earth)

Fig. 3.33 Debris flow around the railway (*left*, after the Mainichi) (A in Fig. 3.32) and flood in the downtown (*right*, after the Kobe Shimbun) (B in Fig. 3.32), July 1938

Fig. 3.34 Landslides in the Rokko Mountains (after Geospatial Information Authority of Japan: GSI) (C in Fig. 3.32), July 1967

used as such. The pond's catchment area covers 28 ha and has a relative relief of 70 m. The average water area is approximately 0.38 ha, and it reaches a maximum depth of approximately 2.8 m. The catchment area is mainly covered with *Pinus densiflora*. Analytical result for the physical property of a sediment core (grain size) is shown in Fig. 3.37a, which indicates two increases in grain size corresponding to these two major natural disasters (Fig. 3.37b; Kashiwaya et al. 2015). 100-mm excess rainfall was employed for the rainfall corresponding to the two events (100-mm excess rainfall is an annual summation of excess amount of 100-mm daily rainfall, which was the criterion for an alarm of heavy rainfall disaster in the district at that time). The tectonic activity and obtained instrumental observations in the Kawauso-ike system will be discussed in the Chaps. 4, 5, 6, and 7.

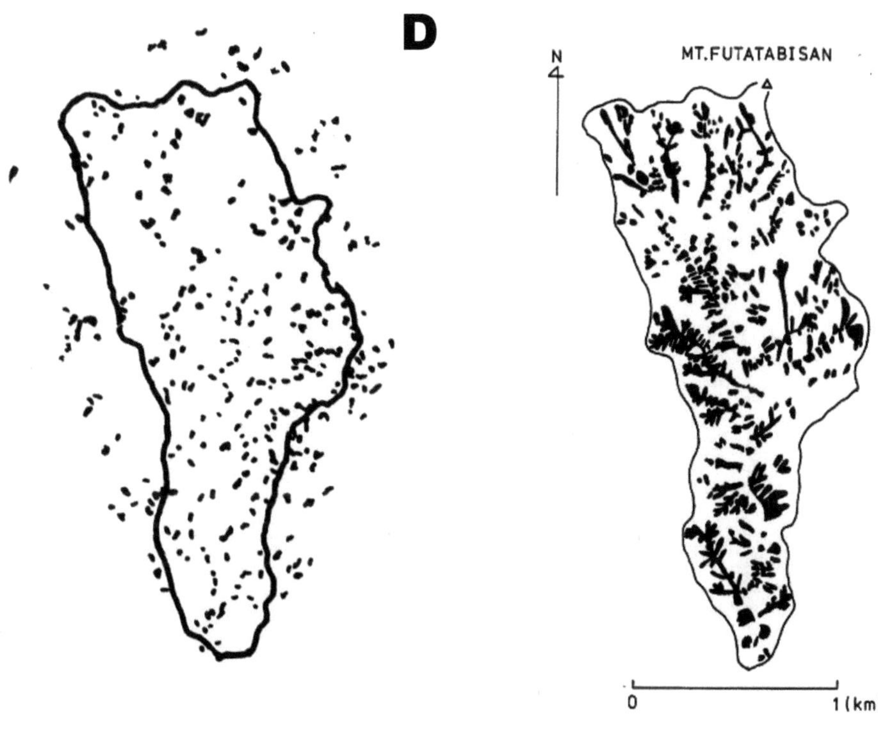

Fig. 3.35 Landslides around the Futatabi catchment of the Rokko Mountains in 1938 (*right*) and 1967 (*left*) (D in Fig. 3.32) (after Kashiwaya et al. 1986)

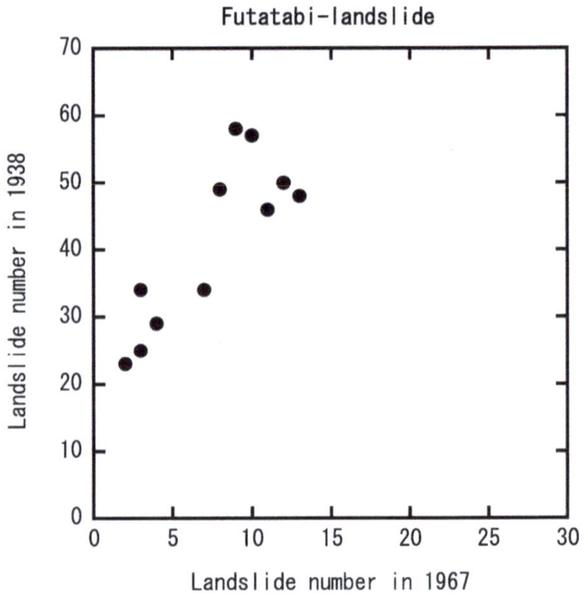

Fig. 3.36 Relationship between landslide number of 1967 per 600 m × 600 m and 1938 in the Futatabi catchment (after Kashiwaya et al. 1986)

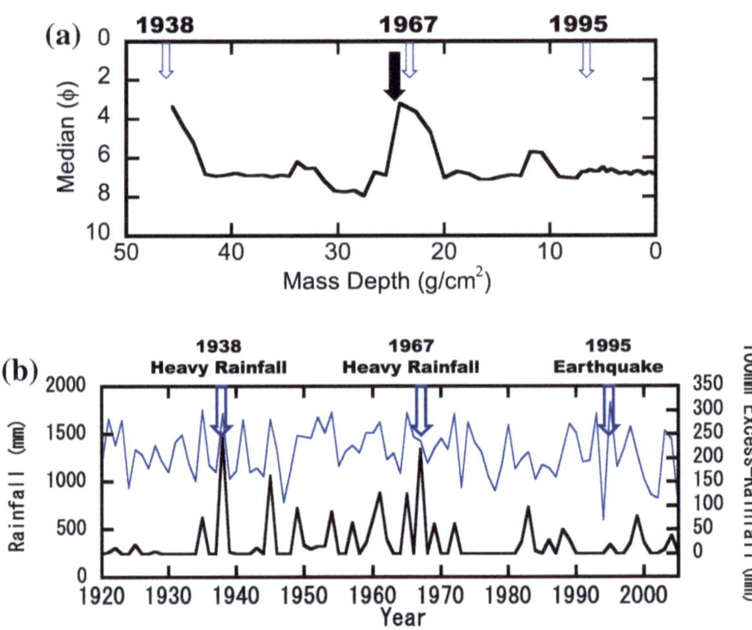

Fig. 3.37 **a** Grain size variation and **b** rainfall (*thin blue line*; annual rainfall, *thick black line*; 100 mm excess rainfall). *Blue arrows* indicate natural disaster events. *Black arrow* means 1963 given by ^{137}Cs concentration

3.3 Abrupt Changes

In lake-catchment systems, the most drastic transformations including large changes in lake and catchment areas due to shifts in climatic regime were often found in the transition interval between glacial and interglacial periods (e.g., Lake Khuvsgul (Hövsgöl) in Mongolia, Fig. 3.12). Generally, large water level changes in lakes significantly affect the sedimentary conditions because they are closely related to erosion and material transportation (Kashiwaya 2012). Dammed lakes related to large water level changes are often formed due to glaciers (glacial lakes) and landslides/debris flows caused by heavy rainfall, earthquakes, etc. They are often broken because the barrier dam bodies (glaciers, massive materials, etc.) are not strong enough.

It is estimated that some large glacial lake outburst floods (GLOFs) in the past had disastrous effects. For example, the Altai GLOF (West Siberia, Russia) of 15 kyr B.P., which was the largest ever experienced (peak discharge = 18×10^6 m^3/s; Baker et al. 1993), formed scablands and left giant bars and ripples downstream and at the bottom bed of a paleolake (Rudoy and Baker 1993; Rudoy 2002; Herget 2005). These can be easily identified even today from space photographs (Fig. 3.38). Another good example of the GLOFs are the Missoula floods (Spokane flood) in Montana, USA (Fig. 3.39), which also occurred in the late Pleistocene (peak discharge = 1.7×10^6 m^3/s; O'Connor and Baker 1992).

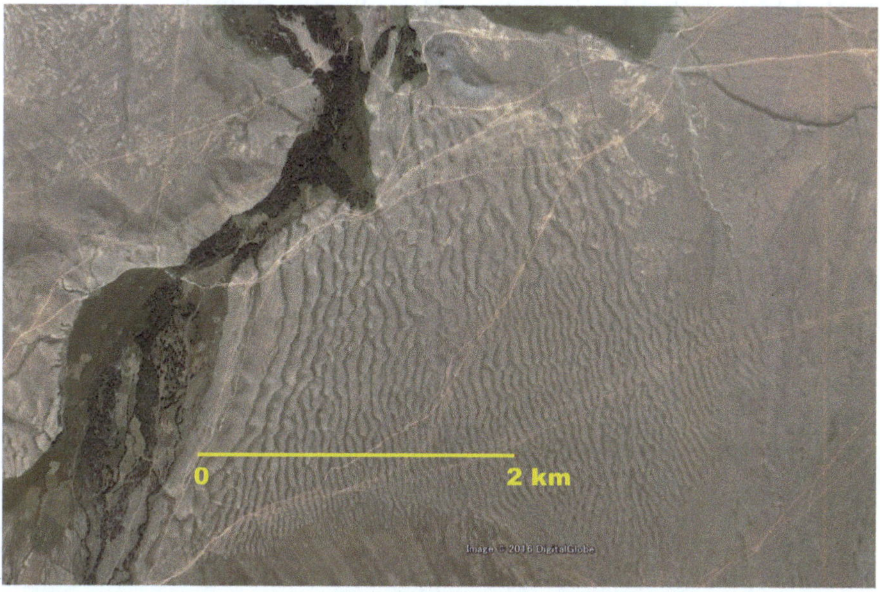

Fig. 3.38 Giant ripples in the Kuray basin, Altai. After Google Earth

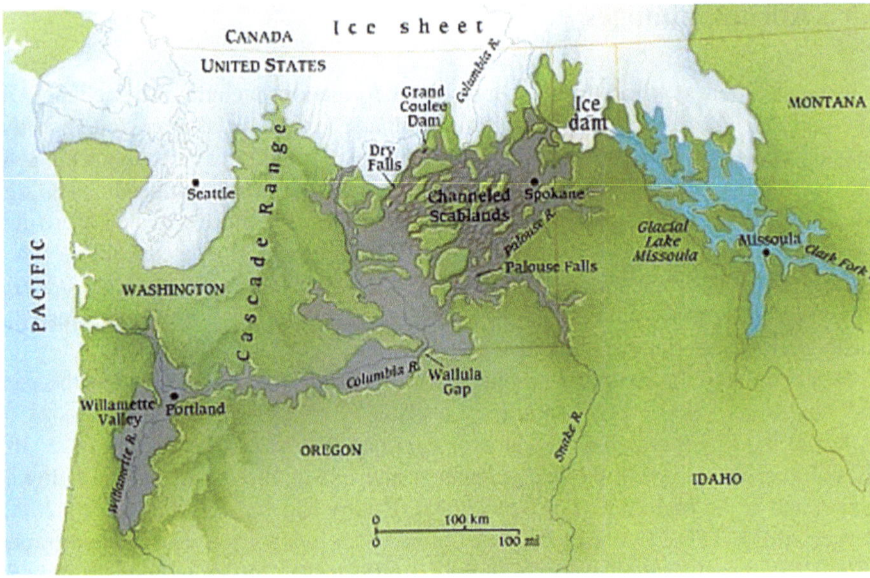

Fig. 3.39 Giant floods in North America. GLOFs from Glacial Lake Missoula (after Waitt 1985)

It is also reported that there were frequent GLOFs around 15 kyr B.P. in the Darkhad basin (paleolake) in northern Mongolia (Fig. 3.40) (Krivonogov et al. 2005; Gillespie et al. 2008; Komatsu et al. 2009), and the peak discharge of the

Fig. 3.40 Location of giant ripples (*left circle*) and glacier barrier of Lake Darkhad (*right circle*) in the upper Yenisey in Tuva, Russia, and Mongolia. Modified after Google Earth

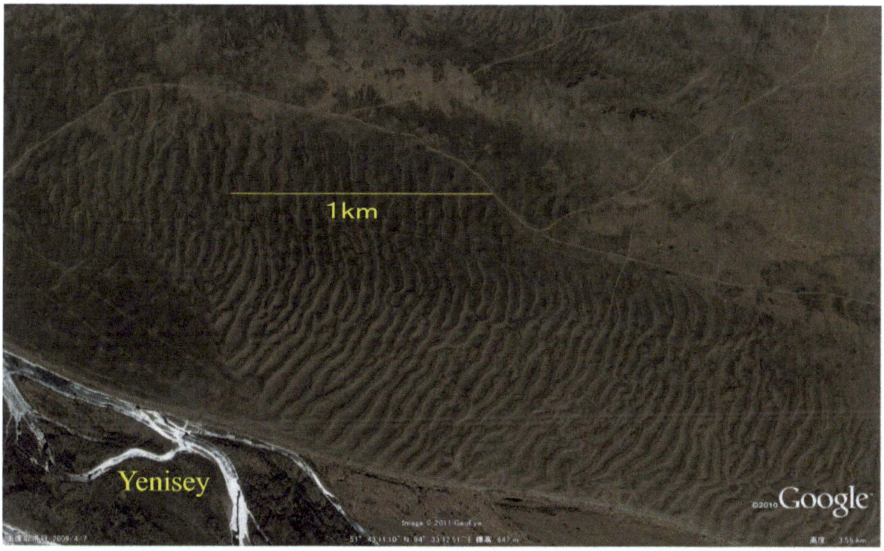

Fig. 3.41 Giant ripples near Kyzyl, Tuva in Russia. Modified after Google Earth

outburst flood from the dammed lake was estimated at 4×10^5 m^3/s (O'Connor and Costa 2004; Rudoy 2002). Probably, giant ripples found in the upper Yenisey (the outflow river of Darkhad) have been related to these GLOFs (Fig. 3.41). Rapid sedimentation in this interval can be found in Lake Darhad sediments (Fig. 3.42 (left), Abe 2011) and Lake Khuvsgul (Hövsgöl; located east of Darhad) sediments (Fig. 3.42 (right), Nakagawa 2010).

Present GLOFs due to dam body burst often occur in the high mountain areas and in the high latitudes, and they are targets of natural disaster control measures in cold regions (e.g., Nepal, Patagonia, Canada, and Scandinavia). Recently, they have often been discussed in the context of global warming (Yamada 2000; Bajracharya and Mool 2009). Some glacial lakes have been monitored for potential GLOFs (ICIMOD 2011; Hambrey et al. 2008; Quincey et al. 2007; Yamada 2000) to understand mechanical processes underlying them and to predict future events. It is also important to obtain information on the temporal changes of the physical conditions relating to GLOFs for natural disaster control. Furthermore, lacustrine sediments recording GLOFs provide valuable information on GLOF processes and changes.

Rainfall-related landslide-dammed lakes are other examples of large external forces (catastrophic heavy rainfall events) at work. In Japan, there have been many landslide-dammed lakes due to heavy rainfall even in the instrumental observation period (since 1875). A famous example is the dammed lake formed in the Totsugawa drainage basin (central Japan) in 1889 (a typhoon attack was presumed: Hirano et al. 1984; Fig. 3.43a). Many landslides occurred (relatively large ones were around 1150 sites), many dammed lakes were formed (53 sites), and

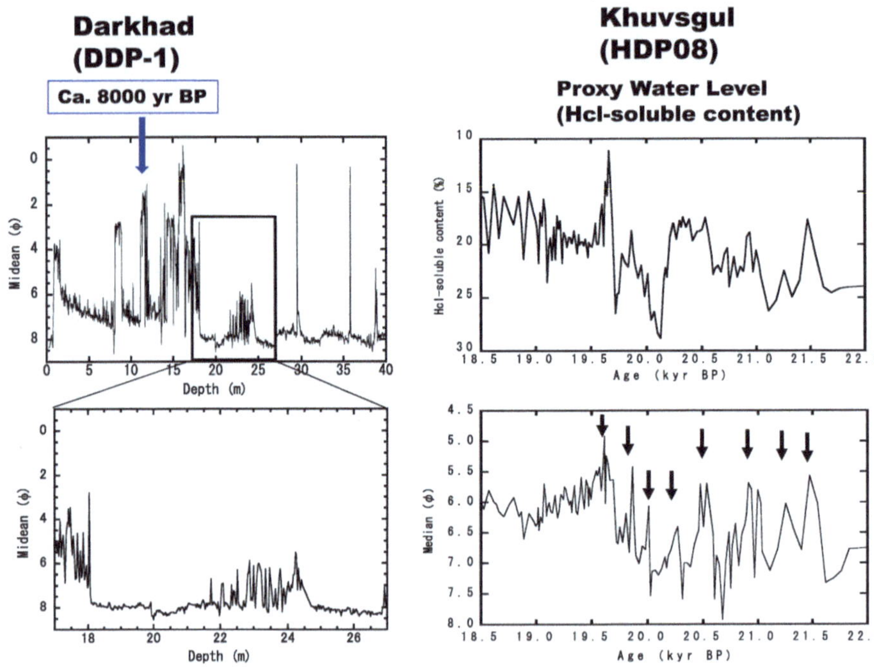

Fig. 3.42 Changes in the mineral grain size in Darkhad (*left*) (Abe 2011) and changes in HCL-soluble content (*upper right*) and mineral grain size (*lower right*) in Khuvsgul (Nakagawa 2010). *Black arrows* mean abrupt increase in grain size

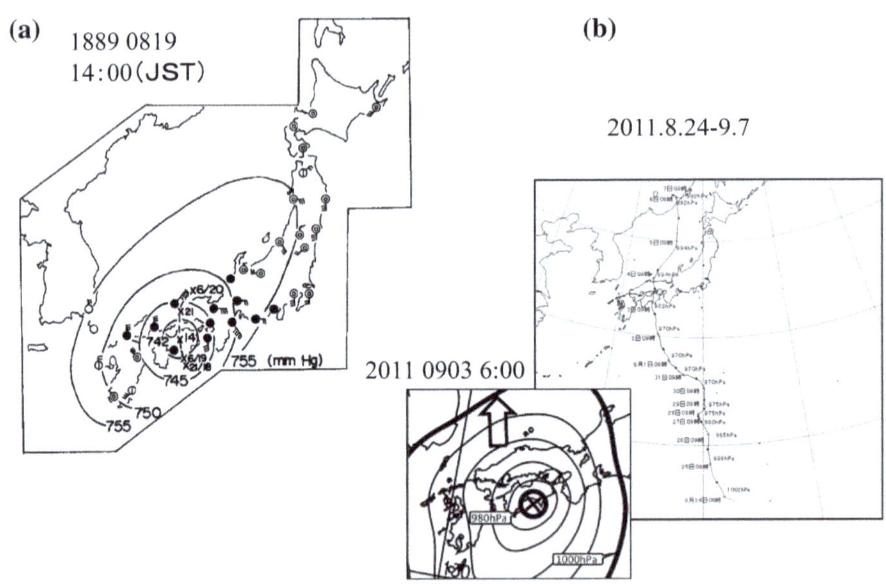

Fig. 3.43 Weather maps and typhoons in **a** 1889, and **b** 2011 around Japan (after Japan Meteorological Agency)

Fig. 3.44 A landslide-dammed lake (Oobatake-doro, Nara, Japan) formed in 1889 (based on GSI MAPs)

approximately 250 people were killed in the drainage basin (Kabata and Kobayashi 2006). These events forced people (more than 2600) in the villages (Totsugawa Go*) to move to Hokkaido, northern Japan. In fact, immigrants developed the land there created the village Shin**-Totsugawa Mura***. Today, a survived dammed lake called Oobatake-doro exists in the disaster area, which is used as a drinking water reservoir (Fig. 3.44). In 2011, another typhoon struck Japan, taking a similar course to the 1889 typhoon (Makihara 2012; Fig. 3.43b), and many landslide-dammed lakes were formed in the Kii Peninsula (central Japan) where Totsugawa Mura is located. After the typhoon, some dammed lakes were at high risk of outburst floods, but they are stable now because of some engineering measures (artificial drainage, etc.).

Most dammed-lake sediments only indicate (the beginning of) disastrous events, and the following ones do not always show continuous records in environmental changes there. However, continuous lacustrine records are required to reveal abrupt events in lake-catchment systems in the context of environmental fluctuations. Records of other lake-catchment systems in the vicinity could provide appropriate interpretation of environmental changes, just as Lake Khuvsgul sediments are of great use for understanding the environmental changes in Lake Darkhad system.

* "Go" means villages, **"shin" means "new", and *** "Mura" means village.

References

Abe H (2011) Hydro-environmental fluctuations inferred from lake sediments in Darhad Paleo-lake (Mongolia). Graduation thesis, Kanazawa University (in Japanese with English abstract)

Aota Y, Kumagai M, Kashiwaya K (2006) Estimation of vertical mixing based on water current monitoring in the hypolimnion of Lake Biwa. JSME Int J Ser B-fluids Therm Eng—JSME INT J SER B. 49(3):621–625

Baker VR, Benito G, Rudoy AN (1993) Palaeohydrology of late Pleistocene super flooding, Altay mountains, Siberia. Science 259:348–350

Bajracharya SR, Mool P (2009) Glaciers, glacial lakes and glacial lake outburst floods in the Mount Everest region, Nepal. Ann Glaciol 50(53):81–86

Biwako chisui-kai (Biwako flood control association) (1925) Biwako chisui enkakusi (Histories on flood control of Lake Biwa). Biwako chisui-kai, Otsu (in Japanese)

Edgington DN, Robbins JA, Colman SM, Orlandini KA, Gustin MP (1996) Uranium-series disequilibrium, sedimentation, diatom frustules, and paleoclimate change in Lake Baikal. Earth Planet Sci Lett 142:29–42

Gillespie AR, Burke RM, Komatsu G, Bayasgalan A (2008) Late Pleistocene glaciers in Darhad Basin, northern Mongolia. Quat Res 69:169–187

Hambrey MJ, Quincey DJ, Glasser NF, Reynolds JM, Richardson SJ, Clemmens S (2008) Sedimentological, geomorphological and dynamic context of debris-mantled glaciers, Mount Everest (Sagarmatha) region, Nepal. Quat Sci Rev 27:2361–2389

Hays JD, Imbrie J, Shackleton N (1976) Variations in the Earth's orbit: pacemaker of the ice ages. Science 194:1121–1132

Herget J (2005) Reconstruction of Pleistocene ice-dammed lake outburst floods in the Altai Mountains, Siberia. Geol Soc Am Spec Pap 386:118p

Hirano M, Suwa H, Ishii T, Fujita T, Gocho Y (1984) Reexamination of Totsugawa hazard in 1889 with special reference to geological control of lange-scale landslides. Annuals, DPRI, Kyoto University 27B-1, pp 369–386

Horie S (ed) (1984) Lake Biwa. Dr. W. Junk Publishers, Dordrecht, 654p

Horie S (ed) (1987) History of Lake Biwa. Kyoto University, Otsu, 242p

ICIMOD (2011) Glacial lakes and glacial lake outburst floods in Nepal. ICIMOD, Kathmandu, Nepal, 99p

IKF (1996) The Baikal map. Irkutsk, Russia

Ishikawa K (2004) Climato-limnological changes inferred from core sediments of Lake Biwa, Japan. Master's thesis, Kanazawa University, 62p (in Japanese with English abstract)

Itono T, Kashiwaya K, Sakaguchi A (2012) Disastrous flood events found in lacustrine sediments from Lake Biwa. Trans Jpn Geomorphol Union 33:453–468

Kabata F, Kobayashi Y (2006) Totsugawa water disaster and emigration to Hokkaido—on the basis of "Records on hydrological disasters at Yoshino-gun in 1889 (Meiji 22 nen Yoshino-gun suisaishi)". Kokon Shoin, Tokyo, 180p (in Japanese)

Kashiwaya K (1987) Theoretical investigation of the time variation of drainage density. Earth Surf Proc Land 12:39–46

Kashiwaya K (2012) Earth surface processes and environmental changes in lake-catchment systems. Trans Jpn Geomorphological Union 33:121–136

Kashiwaya K, Okimura T, Hirano M, Okuda S (1986) Geomorphological aspects of landslide and time change of rainfall character in the south-west part of the Rokko-san Mountains. Annuals DPRI Kyoto Univ. 27B-1, pp 397–408 (in Japanese with English abstract and captions)

Kashiwaya K, Okimura T, Kawatani T (1987) Critical precipitation conditions for landslide and tree ring response in the Rokko Mountains, Kobe, Japan. Int Assoc Hydr Sci 168:191–197

Kashiwaya K, Taishi H, Okimura T, Kawatani T (1988) Grain size variation of pond sediments and landslide environment in the Rokko Mountains. Trans Jpn Geomorphol Union 9:69–77 (in Japanese with English abstract)

Kashiwaya K, Fukuyama K, Yamamoto A (1991) Time variation of coarse materials from lake-bottom sediments and secular paleoclimatic change. Geophys Res Lett 18:1245–1248

Kashiwaya K, Okimura T, Kawatani T, Aoki T, Isozumu Y (1995) Surface erosional environment and pond sediment information. In: Slaymaker O (ed) Steepland Geomorphology. Wiley, Chichester, 283p, pp 219–231

Kashiwaya K, Ochiai S, Sumino G, Tsukamoto T, Szyniszewska A, Yamamoto M, Sakaguchi A, Hasebe N, Sakai H, Watanabe T, Kawai T (2010) Climato-hydrological fluctuations recorded in long lacustrine records in Lake Hövsgöl, Mongolia. Quat Int 219:178–187

Kashiwaya K, Okimura T, Itono T, Ishikawa K, Kusumoto T (2015) Present earth-surface processes and historical environmental changes inferred from lake-catchment systems. In: Kashiwaya K, Shen J, Kim JY (eds) Earth surface processes and environmental changes in East Asia—records from lake-catchment systems. Springer, Tokyo, 321p, pp 1–24

Komatsu G, Arzhannikov SG, Gillespie AR, Burke RM, Miyamoto H, Baker VR (2009) Quaternary paleolake formation and cataclysmic flooding along the upper Yenisei River. Geomorphology 104:143–164

Krivonogov SK, Sheinkman VS, Mistruykov AA (2005) Stages in the development of the Darhad dammed lake (Northern Mongolia) during the Late Pleistocene and Holocene. Quat Int 136:83–94

Makihara Y (2012) Typhoon Talas in 2011 and Totsugawa natural disaster in 1889 (Heisei 23 nen Taifu dai 12 go to 1889 Totsugawa saigai). Tenki 59(3):151–155 (in Japanese)

Milanković M (1920) Théorie mathématique des phénomènes thermiques produits par la radiation solaire. Académie Yougoslave des Sciences et des Arts de Zagreb, Gauthier-Villars

Milanković M (1941) Kanon der Erdbestrahlung und seine Andwendung auf das Eiszeitenproblem. R. Serbian Acad. Belgrade, 633p

Nakagawa T (2010) A study on paleoclimate and abrupt environmental change inferred from sediment analyses of Lake Hovsgol. Graduation thesis, Kanazawa University, 37p (in Japanese with English abstract)

Ochiai S, Kashiwaya K (2005) Climato-hydrological environment inferred from Lake Baikal sediments based on an automatic orbitally tuned age model. J Paleolimnol 33:303–311

Ochiai S, Kashiwaya K (2003) Hydro-geomorphological changes and sedimentation processes printed in sediments from Lake Baikal. In: Kashiwaya K (ed) Long continental records from Lake Baikal. Springer, Tokyo, 370p, pp 297–312

O'Connor JE, Baker VR (1992) Magnitudes and implications of peak discharges from glacial Lake Missoula. Geol Soc Am Bull 104:267–279

O'Connor JE, Costa JE (2004) The world's largest floods, past and present: their causes and magnitudes. U.S. Geological Survey Circular 1254, 13p

Okuda S, Imberger J, Kumagai M (eds) (1995) Physical processes in a large lake: Lake Biwa, Japan. Coastal Estuarine Stud, vol 48. AGU, Washington, 216p

Quincey DJ, Richardson SD, Luckman A, Lucas RM, Reynolds JM, Hambrey MJ, Glasser NF (2007) Early recognition of glacial lake hazards in the Himalaya using remote sensing datasets. Glob Planet Change 56:137–152

Rudoy AN (2002) Glacier-dammed lakes and geological work of glacial superfloods in the late Pleistocene, Southern Siberia, Altai Mountains. Quat Int 87:119–140

Rudoy AN, Baker VR (1993) Sedimentary effects of cataclysmic late Pleistocene glacial outburst flooding, Altay Mountains, Siberia. Sed Geol 85:53–62

Shiga Prefecture (Shiga-ken) (1966) Records of natural disaster in Shiga prefecture (Shiga-ken saigaishi). Otsu, Shiga, 185p (in Japanese)

Shimada T, Kashiwaya K, Masuzawa T, Hyodo M (2002) Hydro-environmental fluctuation in a fake—catchment system during the late Holocene inferred from Lake Yogo sediments. Trans Jpn Geomorphol Union 23:415–431 (in Japanese with English abstract)

Takamatsu T, Takada J, Tanaka A, Kawai T (2003) Environmental changes during the past 5 Ma inferred from element compositions of Lake Baikal sediments (BDP93 and BDP96). Chikyu Mon Spec Issue 42:108–118 (in Japanese)

Tsukamoto T (2005) Environmental changes in drainage basins inferred from sediment information of Lake Hovsgol and Lake Baikal. Master's thesis, Kanazawa University, 34p (in Japanese with English abstract)

Waitt RB Jr (1985) Case for periodic, colossal jökulhlaups from Pleistocene glacial Lake Missoula. Geol Soc Am Bull 96:1271–1286

Yamada T (2000) Glacier lake outburst floods in Nepal. Seppyo 62:137–147 (in Japanese with English abstract)

Yamamoto Y, Kashiwaya K, Fukuyama K (1984) Periodic variations of grain size in the 200 meter core samples from Lake Biwa. Trans Japan Geomorphol Union 5:345–352 (in Japanese with English abstract)

Yumoto M (2010) Environmental changes in a drainage basin system inferred from sediment information of Lake Hovsgol. Graduation thesis, Kanazawa University, 26p (in Japanese with English abstract)

Chapter 4
Tectonic (Tectono-Geomorphic) Forces on Systems

There are several cases of tectono-geomorphic forces prompting changes in lake-catchment systems, such as Lake Biwa and Lake Baikal systems. The basic structure of the present system around Lake Biwa was established at approximately 450 kyr B.P. (Meyers et al. 1993) mainly due to the uplift of the Hiei Mountains and Hira Mountains located west of the lake and the lake's subsidence (Yokoyama 1984). In fact, there are many faults west of the lake, which are related to the tectonic movement. Examining the analytical results for the upper silt–clay part (250 m; 450 kyr) shows that rapid decrease in the mineral grain size (an indicator of discharge) during the initial sedimentation phase and gradual increase since then may reflect the influence of comparatively rapid subsidence and gradual aggradation of the lake, suggesting a change in water level. On the contrary, a small gradual decrease is detected in the mineral content fluctuation, although the periods corresponding to interglacial ones are clearly decreased (Fig. 4.1).

Many earthquakes even in historical period have occurred in lake-catchment systems. The most disastrous earthquakes ever documented in Lake Biwa system were the 1185 Genryaku (Bunji) earthquake and the 1662 Kanbun earthquake (Usami 2003). Liquefaction and flooding due to the Genryaku earthquake were clearly imprinted in archeological records of the old port town (Shiozu port) at Shiozu Bay in the north Lake Biwa near Lake Yogo (Yokota 2011). Some earth-surface processes during the same period as the earthquake may have triggered a small piracy in the northwest mountains (Fig. 4.2), where the gradual hydrological headward erosion of the Momose River also played an important role (Mizuyama et al. 1975). After the piracy, a small pond called Daira-ike was formed, and the age of the deepest bottom sediment was dated approximately 700 yr BP (Matsui 1988), which supports that the piracy occurred around the period of the Genryaku earthquake in 1185. Its source faults were estimated at the south part of the Biwako-seigan (Lake Biwa western coast) fault zone (black dotted line in Fig. 4.3) (Komatsubara 2012). In the previous section (Fig. 3.17), the ages of the Genryaku earthquake and the Kanbun earthquake were estimated in the T-1 core (Genryaku; green arrow in Fig. 4.4a) obtained in Shiozu Bay of Lake Biwa near Lake Yogo (Fig. 4.3). The same trend is

K. Kashiwaya, *Geomorphology of Lake-Catchment Systems*,
Environmental Earth Sciences, DOI 10.1007/978-981-10-5110-4_4

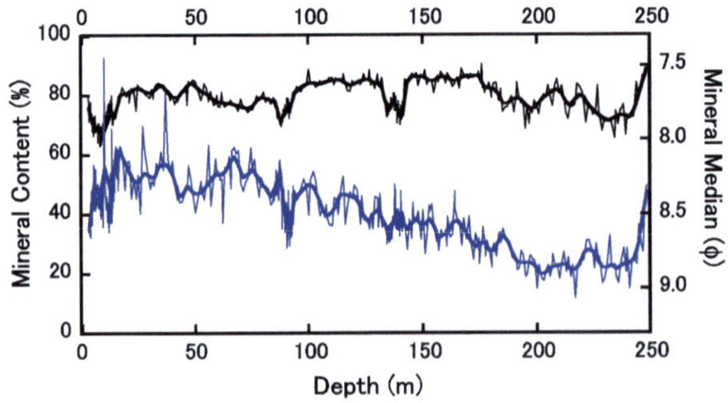

Fig. 4.1 Changes in the mineral content (*black*) and mineral median (*blue*) of lacustrine sediments (*thin*; original, *thick*; filtered) from Lake Biwa (stacked data for 200-m core and 1400-m core; original data given by Ishikawa 2004)

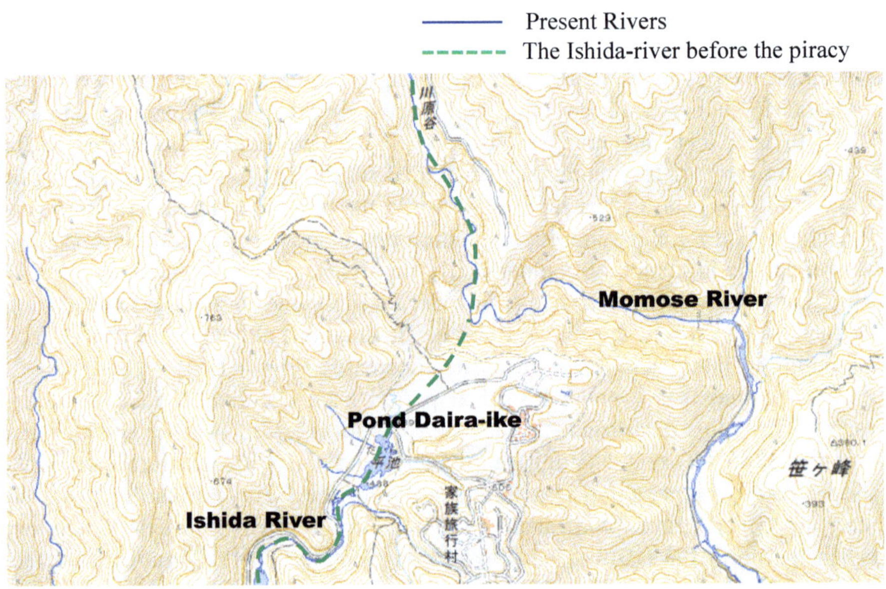

Fig. 4.2 A piracy in the northwest of Lake Biwa catchment at around the 1185 Genryaku earthquake (modified after GSI MAP)

observed around the same period in the physical properties of the sediments from Lake Yogo core (YG3) (Genryaku; green arrow in Fig. 4.4b). Two different physical properties (blue; mineral content or density) in addition to the mineral grain size (red) are used for the two cores in this figure because the pattern of the mineral content is nearly the same to that of density in most cases.

The Kanbun earthquake occurred in 1662 (Usami 2003) at the central-west Lake Biwa and resulted the death of approximately 900 people in the areas around Lake

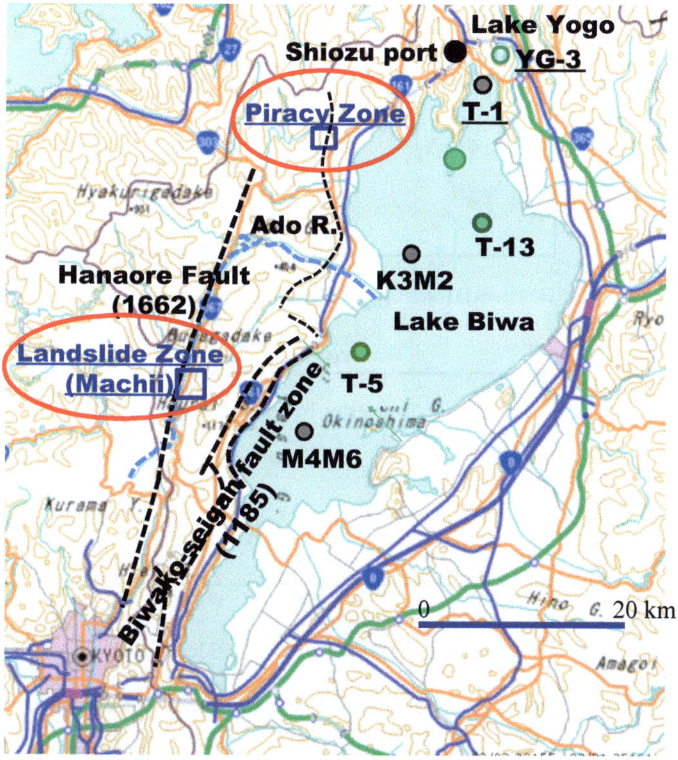

Fig. 4.3 Core sampling points in Lake Biwa and Yogo (*circles*) and tectono-geomorphic zones (*rectangle*). *Dotted curves* indicate active fault lines related to the 1185 earthquake (Biwako-seigan fault) and the 1662 earthquake (Hanaore Fault) (modified after GSI MAP)

Biwa and Kyoto (the capital of Japan at the time) (Nishiyama and Komatsubara 2006). The earthquake was attributed to the activation of Hanaore Fault (Komatsubara 2006) (black dotted line in Fig. 4.3). In the catchment of Lake Biwa, many landslides and liquefactions occurred (Nishiyama 2006). In particular, a large landslide in the western catchment (Machii) killed approximately 560 people and dammed the river Katsura-gawa (Ado River) (Fig. 4.5; Iwamura et al. 2002). Then, the dammed lake burst out after two weeks and significant sediment discharged into Lake Biwa through the Ado River. In Lake Biwa, various core samples were obtained in several points also by the joint team of Kanazawa University and Lake Biwa Research Institute (Fig. 4.3) (Kashiwaya et al. 2015). Analytical results for core sediments obtained at three points in the lake (T-1, K3M2, and M4M6 in Fig. 4.3) are shown in Fig. 4.6 (Kashiwaya et al. 2015), indicating that the mineral grain size decreased for all sampling points, whereas the mineral content increased around the earthquake year (1662; green dotted line in Fig. 4.6). A similar decrease in the mineral grain size and increase in the grain density (mineral content) are also detected in Lake Yogo sediments (YG4 in Fig. 3.18, YG3 in Fig. 4.4). These may indicate that the shaking of the

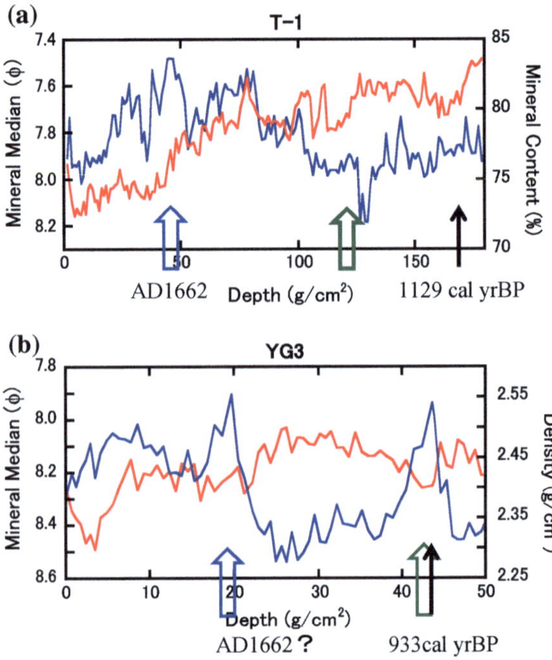

Fig. 4.4 **a** Changes in mineral grain size (median) (*red*) and mineral content (*blue*) of T−1 core in Lake Biwa (Kashiwaya et al. 2015), and **b** changes in mineral grain size (median) (*red*) and density (*blue*) of YG3 core in Lake Yogo (Shimada et al. 2002). *Blue arrows* mean estimated year of 1662, *green* ones mean estimated year of 1185 and *black* ones mean ^{14}C dates

earth surface resulted in the release of a large volume of material (landslides, etc.), and fine materials were easily transported into lakes even during small rainfalls (Kashiwaya et al. 2004, 2015). Possibly, the erosion and/or transportation conditions in the catchment may change owing to large earthquakes.

Earth-surface processes due to tectonic activity are also imprinted in the Baikal sediments although it is almost impossible to detect the details because of the comparatively low resolution (low sedimentation rate) in the sediments (except river mouth sediments), as compared to Biwa sediments. This area has experienced significant tectonic activity because it belongs to the Baikal rift zone. Changes in these processes are generally reflected on the catchment conditions. It is hypothesized that two major tectonic events affected the area around Lake Baikal catchment during the past 1.0 M yrs (Mats and Perepelova 2011): the Primorsky tectonic event (1.0–0.8 Ma) and the Tyya tectonic event (two glacials, MIS6 and MIS8, are candidates for this event although the interval 0.15–0.12 Ma was assumed for this event, which corresponds to MIS6). Listvyanka block subsidence (0.06 Ma) is also picked up as a post-tectonic event. The Primorsky event corresponded to tectonic uplift in the Primorsky Mountains, west of Lake Baikal, and caused a switch of the flow direction of the uppermost part of the Rena River. In fact, a tributary of the

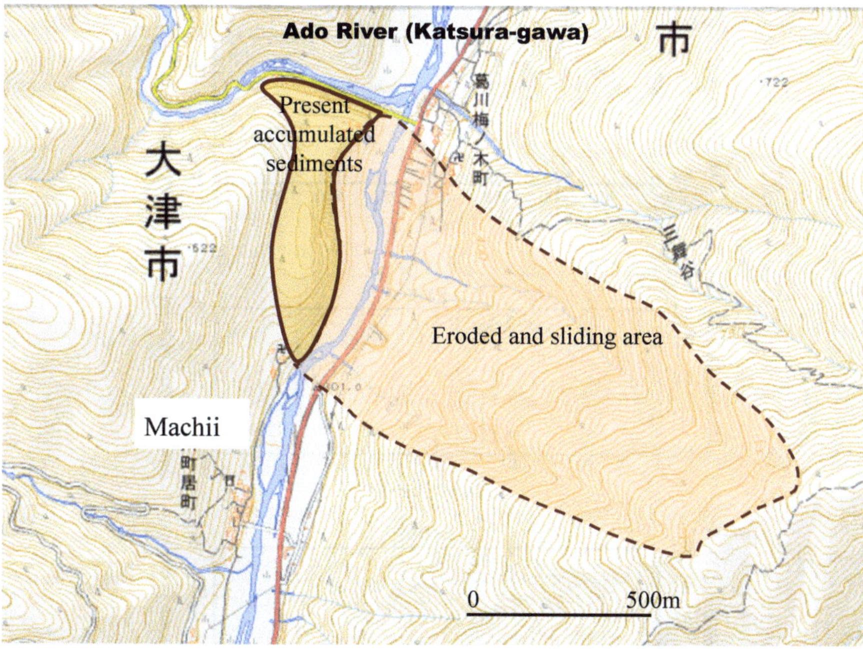

Fig. 4.5 A big landslide at Machii in the 1662 Kanbun earthquake (modified after GSI MAP)

Rena that originally flowed out from Lake Baikal became a tributary of the Buguldeyka River, and it now flows into Lake Baikal (Fig. 4.7). Before this tectonic event, the Baikal was drained out through the paleo-Manzurka River into the Rena River. After the event, the water level rose and water flowed out into the Yenisei River through a new drainage system (the Irkut–Ilcha valley) (Fig. 4.8).

Bezrukova et al. (2004) assumed that the hiatus around 134 m in the BDP99 core corresponded to the Primorsky event. The core was obtained from the Posolskaya Bank in front of the delta area of the Selenga river mouth (Fig. 3.6) (52° 05′23″ N, 105°15′24″ E). Analytical results for the core sediments show that the sediments are mostly mineral; the mineral grain size and the whole grain size are nearly the same, and the grain size is a proxy for sediment discharge and/or precipitation in the catchment (Fig. 4.9). Two shifts were identified in the stratigraphic sequence based on tracing acoustic reflectors at depths of 92.5 and 44.5 m (BDP99 members 2005). It was reported that Lake Baikal became a deep lake bounded by high flanking mountains during the Tyya tectonic event (Mats and Perepelova 2011). A decrease in the mineral grain size was detected in the upper part of the BDP99 core at a depth of approximately 45 m, and the grain size became comparatively small upward, suggesting deepwater conditions above that depth (after a period of ca 300 ka) (Fig. 4.9). The analytical results (BDP98) from the Academician Ridge also indicate comparatively small grain size at the same interval (Fig. 4.10), supporting the hypothesis that the water level increased. The influence of the Tyya tectonic event may have started during that time (MIS8). Moreover, the

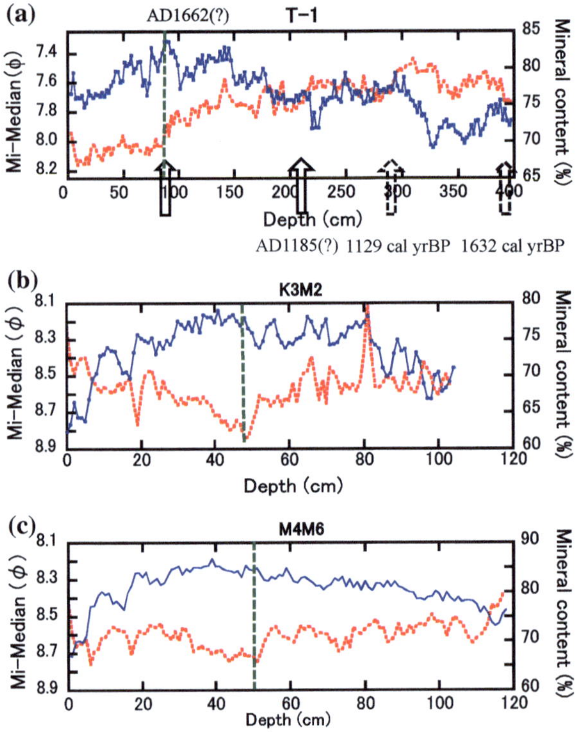

Fig. 4.6 Changes in mineral grain size (median) (*red*) and mineral content (*blue*) of **a** T−1 core, **b** K3M2 core, and **c** M4M6 core in Lake Biwa. *Solid arrows*; ^{14}C estimated date, *dotted arrows*; ^{14}C date, and *dotted green vertical lines*; 1662 (estimated) (modified after Kashiwaya et al. 2015)

water level change indicated by the combination of several chemical properties in the BDP93 core obtained at the Buguldeyka Saddle (offshore the river Buguldeyka) (Takamatsu et al. 2003) supports the claim that the water level started to increase already since 0.28–0.30 Ma (Fig. 4.11).

Typically, tectono-geomorphic forces affecting lake-catchment systems in Japan are volcanic and seismic in origin. As mentioned above, landslides and debris flows have often occurred in association with volcanic activity (1640 Hokkaido Komagatake eruption; 1888 Bandai-san eruption; 1926 Yakedake eruption, etc.) and large earthquake activity (1923 Kanto Great earthquake, 1995 Kobe earthquake, 2011 East Japan Great earthquake, etc.). In some cases, landslides and debris flows dammed valleys and rivers, and new lake-catchment systems were established.

The present Lake Onuma system in Hokkaido (northern Japan) was established in 1640 by mountain decay with pyroclastic flows in the eruption of Mt. (Hokkaido) Komagatake (Yoshimoto and Ui 1998). Then, the Orito-gawa, south of the mountain, was dammed by the flows (Fig. 4.12), and the present Onuma lake-catchment system was established. The flows also entered the sea (the Pacific Ocean) and produced a giant tsunami, which resulted in more than 700 deaths in the

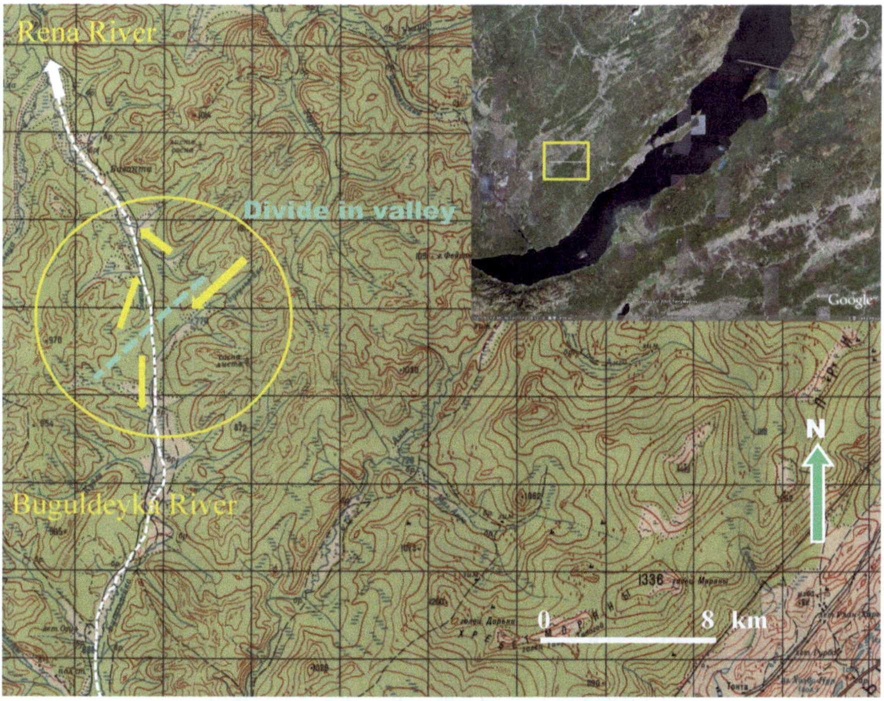

Fig. 4.7 Divide in valley in the area of the upper Rena and the upper Buguldeyka (a side of a grid; 4 km) (modified after Kashiwaya 2012; original figures by Google Earth and IKF 1996)

Fig. 4.8 Location of the Irkut–Ilcha valley (ellipse) (modified after FSGCR 2002)

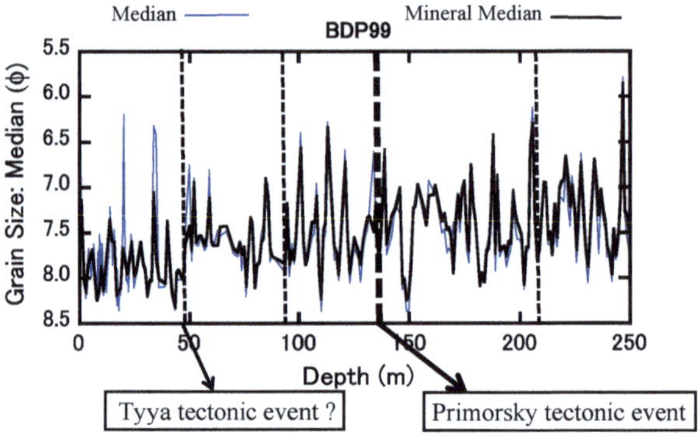

Fig. 4.9 Changes in sediment median (*thin blue curve*) and mineral median (*bold black curve*) of the BDP99 core. *Dotted lines* mean boundaries given by acoustic signal

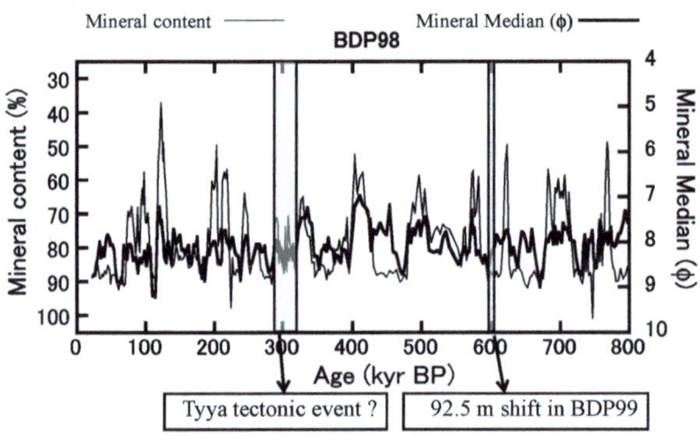

Fig. 4.10 Changes in mineral content (*thin curve*) and mineral median (*bold curve*) of BDP98 core

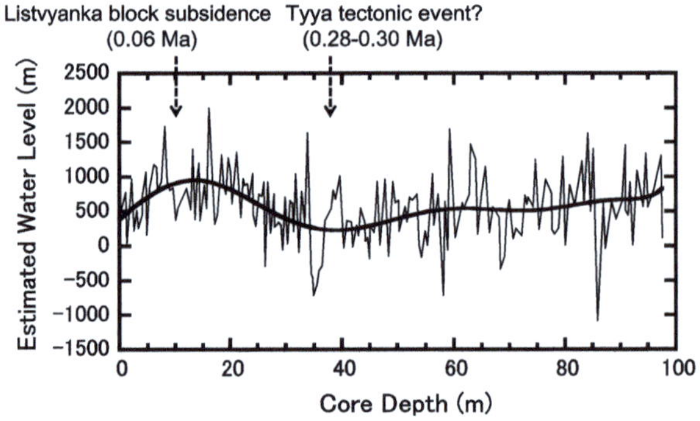

Fig. 4.11 Estimated water level change in Lake Baikal (modified after Takamatsu et al. 2003)

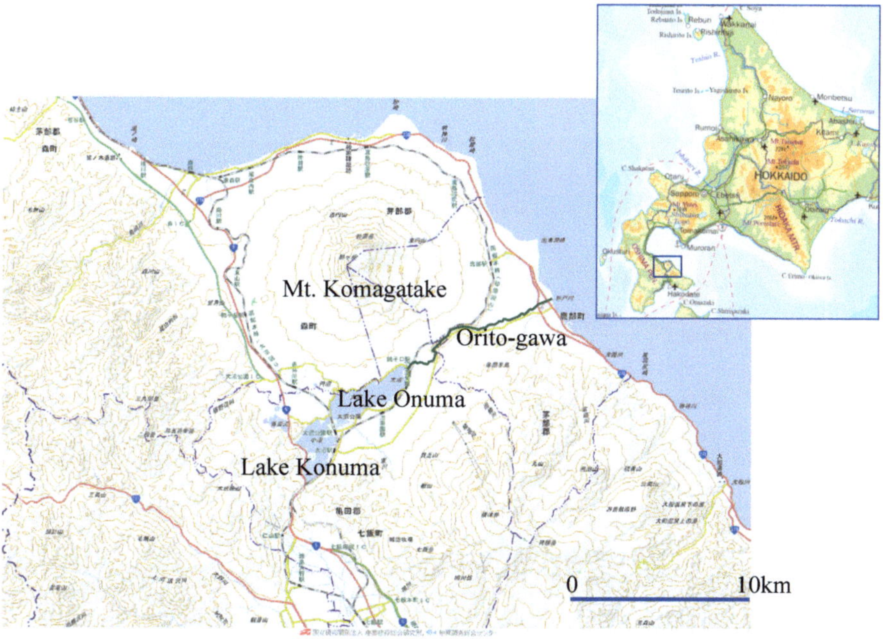

Fig. 4.12 Mt. Komagatake and Lake Onuma system in Hokkaido, Japan (based on GSI MAPS)

Table 4.1 Recent eruptions of Mt. Komagatake in Hokkaido, Japan (modified after Japan Meteorological Agency and Volcanological Society of Japan 2013)

Age	Size	Remark
2000 (Sept., Oct.)	Small VEI 1	Phreatic eruption
1998 (Oct.)	Small VEI 1	Phreatic eruption
1996	Small VEI 1	Phreatic eruption
1942	Medium VEI 2	Phreatic eruption → phreatomagmatic eruption → phreatic eruption, pyroclastic surge, lahar
1937	Small	Phreatic eruption
1929	Large VEI 4	Magmatic eruptions, pyroclastic flow, Ko-a tephra
1924	Small	Phreatic eruption
1923	Small	Phreatic eruption
1919	Small	Phreatic eruption
1905	Small	Phreatic eruption, secondary lahar, Ko-b tephra
1888	Small	Phreatic eruption
1856	Large VEI 4	Magmatic eruptions, pyroclastic flow, Ko-c1 tephra
1765	Small ?	Ancient legend
1694	Large VEI 4	Magmatic eruption, pyroclastic flow, Ko-c2 tephra
1640	Large VEI 5	Magmatic eruption, sector collapse, debris avalanche, pyroclastic flow, Ko-d tephra

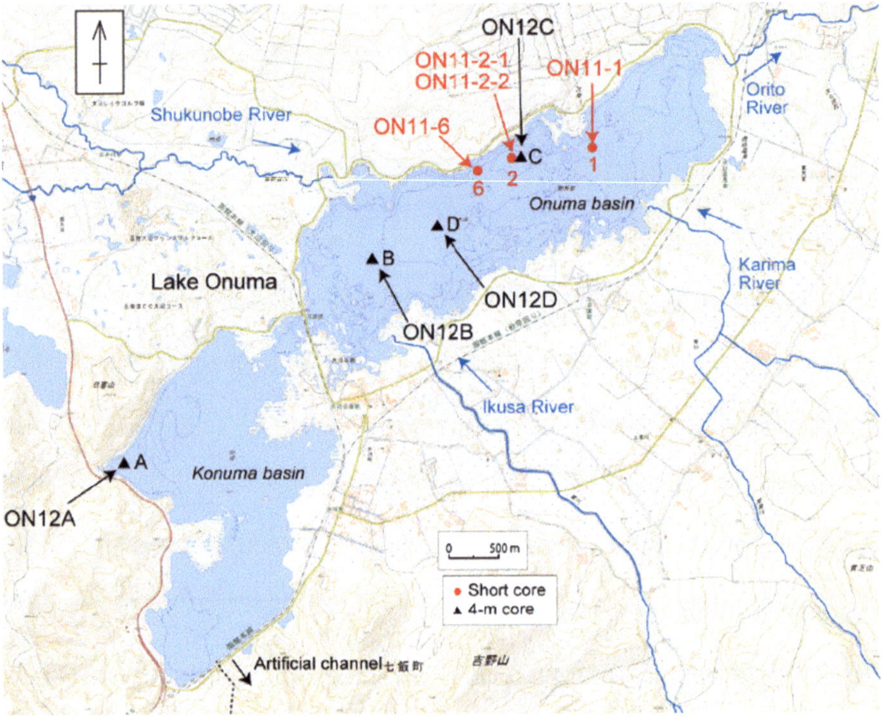

Fig. 4.13 Sampling points in Onuma system. *Alphabets* (*black*, *triangles*) mean long cores and *solid circles* (*red*) mean short cores (based on GSI MAPS) (Itono et al. 2015)

opposite coast. Since the 1640 eruption, there have been four large eruptions (Table 4.1), and the lacustrine sediments include precious information on the late Little Ice Age (LIA) and the post-LIA period. Three long cores and several short cores were obtained in the Onuma lakes (Lake Onuma and Lake Konuma) in 2012 (Fig. 4.13), and some analytical results have been already published (Hasebe et al. 2015; Itono et al. 2015; Ochiai et al. 2015). The descriptions and the water content in the long cores are shown in Fig. 4.14 (Hasebe et al. 2015). The varve structure after the 1694 eruption in this figure corresponds to an interval without bioturbation (or with little bioturbation), which could be interpreted as a cold period. Sun et al. (2016), based on the $\delta^{15}N$ records in the long core (ON12C), assumed that the precipitation increased after the eruption and until the end of the nineteenth century. The large eruption in 1929 was also recorded in the short cores (Figs. 5.10 and 5.11), which offers insight into the effect of human activity (see Chap. 5).

The eruption of Mt. Yakedake in central Japan in 1915 dammed the Azusa River with volcanic mudflow and formed a pond called Taisho-ike (Fig. 4.15). Since then, several eruptions and debris flows occurred catchment since 1970 (Murai 1962; Oikawa et al. 2002). Field observation of the debris flows in the catchment (main target for the observation was a gully called Kamikami-horisawa gully) was a pioneering disaster

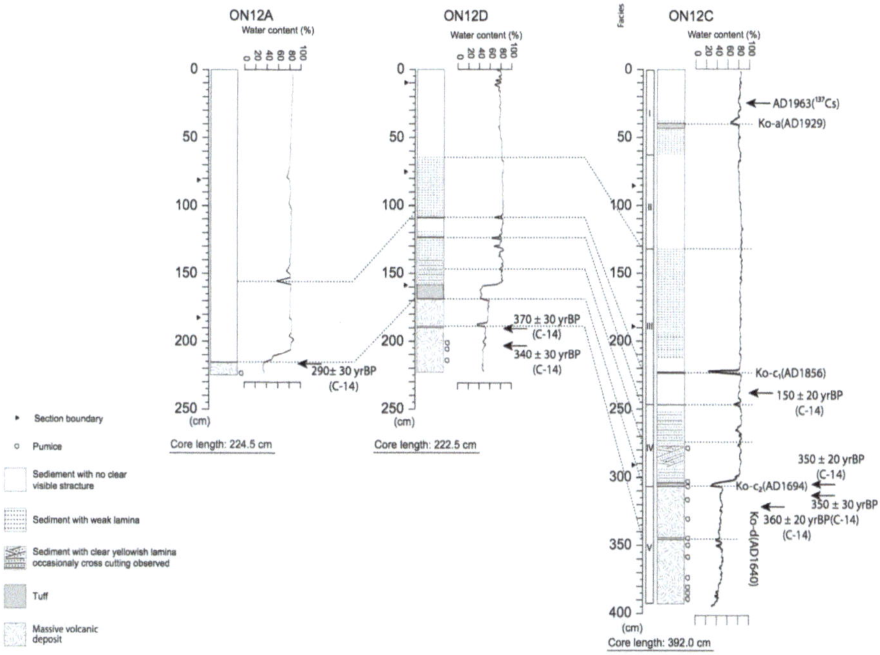

Fig. 4.14 Description and water content of three long cores (Hasebe et al. 2015)

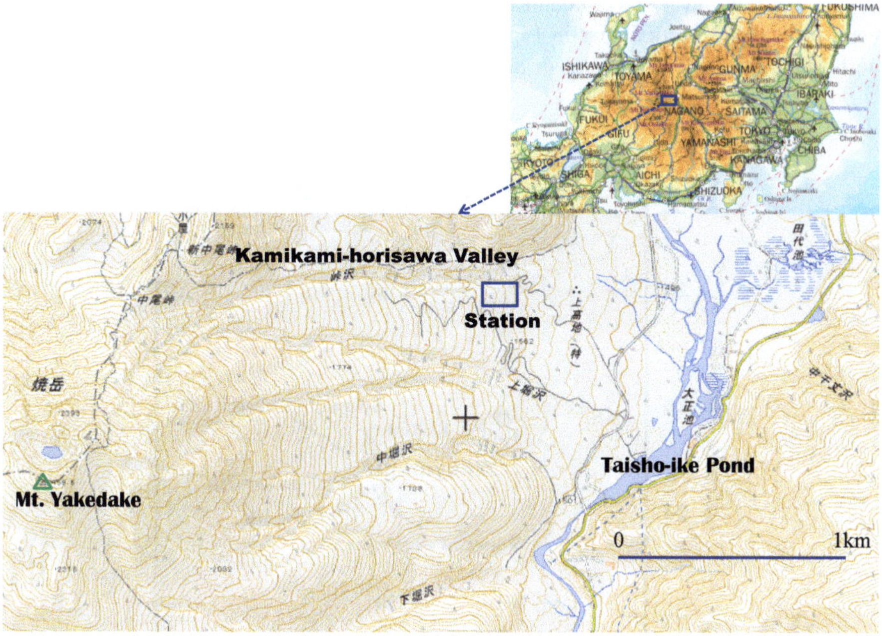

Fig. 4.15 Debris flow station in Mt. Yakedake, central Japan (based on GSI MAPs)

prevention study (the produced video of debris flows was probably one of the first records of this phenomenon in the world) (Okuda et al. 1980; Suwa 1988). However, sediments due to the debris flows were often artificially dredged to preserve the scenery around Taisho-ike (this area is located in a national park); hence the information on the lacustrine sediments could not be connected to the recent debris flows.

Some disastrous earthquakes have occurred in China recently (the 1976 Tangshan earthquake M_w7.5, the 2008 Wenchuan earthquake M_w7.9, etc.). For example, the 1970 Tonghai earthquake M_s7.7 in Yunnan (Fig. 4.16) killed more than 15000 people. As this earthquake took place in the Cultural Revolution period (1966–1976), some local people thought that it might be an attack with atomic bombs by the former Soviet Union (Asahi Shimbun 2001). In addition, the 1833 Songming earthquake M8.0 happened in 1833 caused substantial damages (Fig. 4.16). These earthquakes happened along the Xiaojiang Fault in the southeast Yunnan Province, as a result of the northward motion of the Indian plate, westward subduction of the Pacific plate, and northwestward impact of the Philippine Sea plate (Yang et al. 2005). Several core samples were obtained in some lakes near Kunming (Fig. 4.17 and Table 4.2); Lake Qilu-hu (Tonghai), Lake Xinyung-hu, and Lake Fuxian-hu in 2000–2002 by a joint team of Kanazawa University and the Yunnan Institute of Geography (Kashiwaya et al. 2006; original data given by Kusumoto 2003).

The lakes are located within 50 km from the epicenter of the Tonghai earthquake, whereas the epicenter of the Songming earthquake was located 90 km north of Fuxian-hu. Thus, the Tonghai earthquake event was recorded in the physical properties of the lacustrine sediments in the lake nearest to the epicenter (Qilu-hu). In fact, the peak in ^{137}Cs concentration (blue circles) of the core was assigned to the year 1963 (blue vertical bar in Fig. 4.18a). The Tonghai earthquake was recorded with local changes in density, grain size, ^{210}Pb concentration, mineral content, and

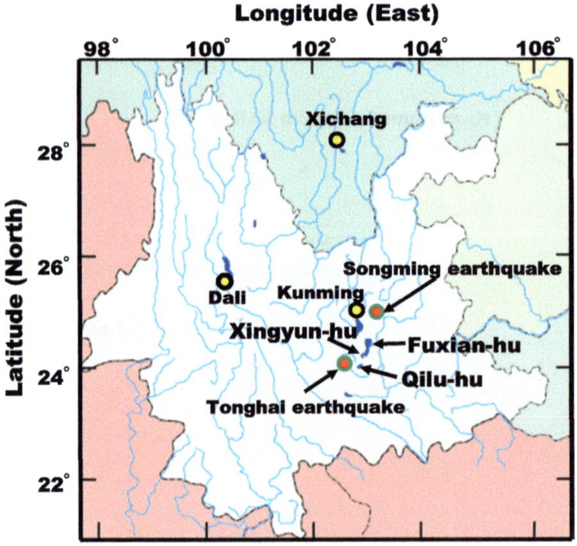

Fig. 4.16 Three lakes studied and two epicenters of recent disastrous earthquakes in Yunnan, China

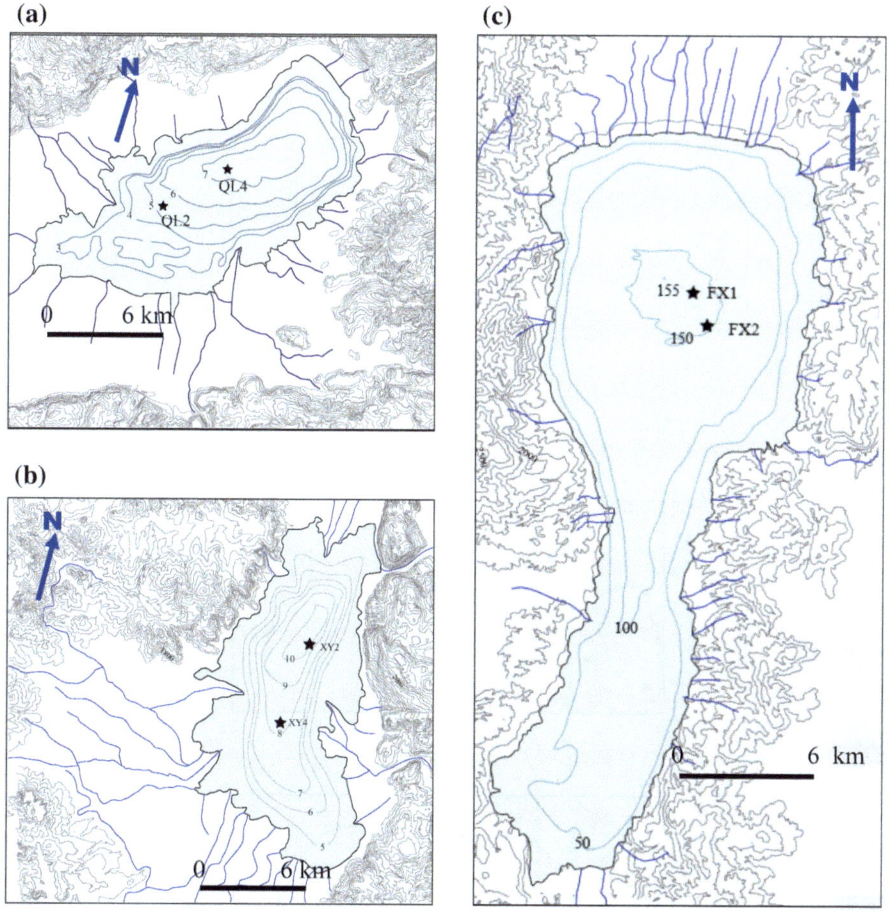

Fig. 4.17 Outline of the three lakes and sampling points; **a** Qilu-hu, **b** Xingyun-hu, and **c** Fuxian-hu

HCl-soluble content in the QL-2 core sediments (light green vertical bar in Fig. 4.18). The dry interval of 1979–1983 (water level fall by 4 m) was also identified (Wan and Dou 1998; gray part in Fig. 4.18), assuming that the age model given by the ^{137}Cs dated peak was accurate. A similar trend is also observed in the upper part of the fluctuations in the central core (QL-4) above the ^{137}Cs peak (Fig. 4.19), although the earthquake event and the dry event are not identified so clearly. This may be related to the fact that the QL-2 core was taken in the shallower point, whereas QL-4 was in the deeper point.

Analytical results for Lake Xinyung-hu, located in the north of Lake Qilu-hu, are shown in Figs. 4.20 and 4.21. They indicate that the earthquake event (Tonghai) is slightly seen in the fluctuations of the central core (XY-2; green vertical bar), assuming a constant sedimentation rate given with ^{137}Cs peak (blue vertical bar). The assumed sedimentation rate suggests that Songming earthquake may correspond to the

Table 4.2 Outline of three lakes in Yunnan (modified after Wan and Dou 1998)

	Qilu-hu	Xingyun-hu	Fuxian-hu
Location	N24°08′ ~ 24°13′ E102°43′ ~ 102°49′	N24°17′ ~ 24°23′ E102°45′ ~ 102°48′	N24°21′ ~ 24°38′ E102°49′ ~ 102°57′
Altitude (m)	1796.8	1722.0	1721.0
Drainage area (km^2)	341	378	1084
Water area (km^2)	36.9	34.7	211.0
Maximum depth (m)	6.8	11.0	155.0
Average depth (m)	4.0	5.3	89.6
C-L ratio	9.2	10.9	5.1

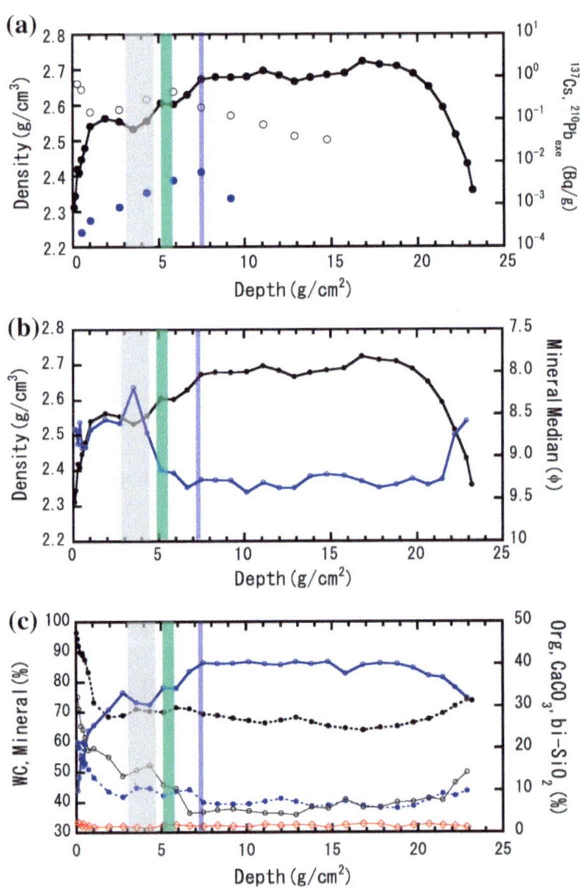

Fig. 4.18 Changes in analytical items for QL-2: **a** density; *black line*, ^{210}Pb concentration; *open circles*, and ^{137}Cs concentration; *blue circles*, **b** density; *black line*, and mineral median; *blue line*, and **c** water content (%); *dotted black line*, mineral content (%); *blue line*, organic content (%); *dotted blue line*, HCl-soluble content (%); *black line*, bi-SiO_2 content (%); *red line*. *Gray vertical bar*; dry interval, *green vertical bar*; Tonghai earthquake, *blue vertical bar*; 1963

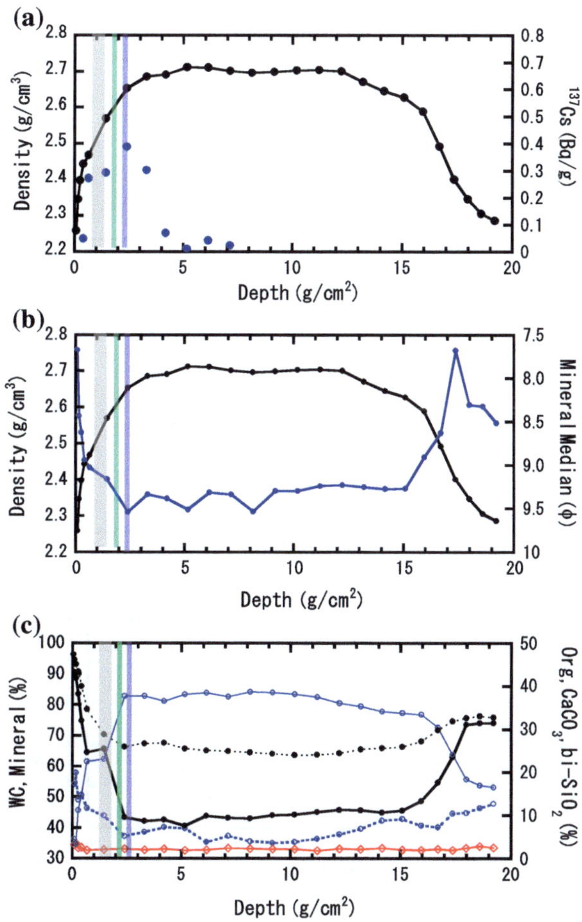

Fig. 4.19 Changes in analytical items for QL-4: **a** density; *black line*, and ^{137}Cs concentration; *blue circles*, **b** density; *black line*, and mineral median; *blue line*, and **c** water content (%); *dotted black line*, mineral content (%); *blue line*, organic content (%); *dotted blue line*, HCl-soluble content (%); *black line*, bi-SiO_2 content (%); *red line*. *Gray vertical bar*; dry interval, *green vertical bar*; Tonghai earthquake, blue vertical bar; 1963

light gray belt. It seems that the two earthquakes are recorded in the sediments (green bar and light gray bar) of the southern core (XY-4), although the age model is estimated through correlation with the XY-2 core. Lake Fuxian-hu is a deep lake (maximum depth = 155.0 m, average depth = 89.6 m), connected to Lake Xinyung-hu through the river called the Ge-he in the south. Analytical results for the cores from this lake are shown in Figs. 4.22 and 4.23, suggesting that the density and the mineral grain size were slightly affected by the Tonghai earthquake in FX-1 core. However, there are comparatively large shifts (green bars) around the earthquake in the FX-2 core, although no exact age model is given for this core. Assuming a constant sedimentation rate for FX-1, the Songming earthquake may correspond to the

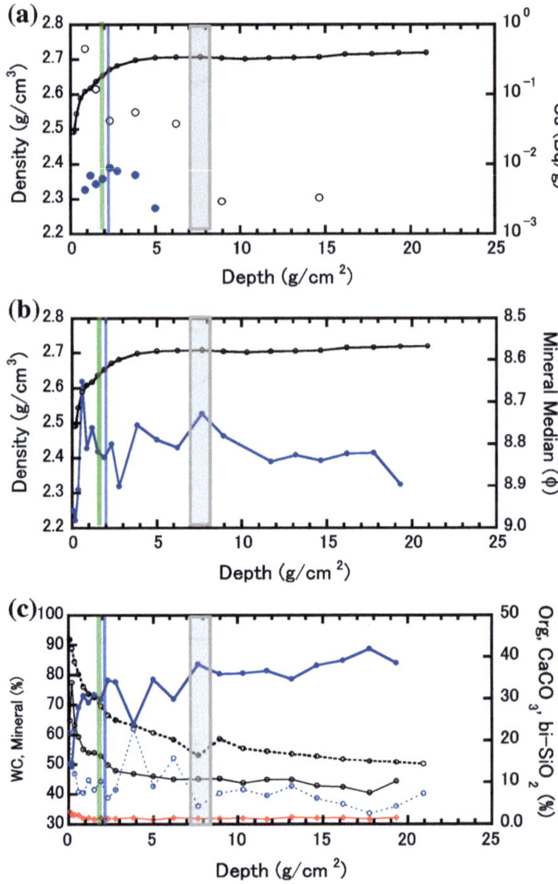

Fig. 4.20 Changes in analytical items for XY-2: **a** density; *black line*, ^{210}Pb concentration; *open circles*, and ^{137}Cs concentration; *blue circles*, **b** density; *black line*, and mineral median; *blue line*, and **c** water content (%); *dotted black line*, mineral content (%); *blue line*, organic content (%); *dotted blue line*, HCl-soluble content (%); *black line*, bi-SiO_2 content (%); *red line*. *Green vertical bar*; Tonghai earthquake, *blue vertical bar*; 1963, and *light gray vertical bar*; Songming earthquake

light gray vertical bar (increase in density around 14–16 g/cm^2) in Fig. 4.22. Additionally, abrupt increases at around the depth of 30 g/cm^2 and 57 g/cm^2 in Fig. 4.22 and at around the depth of 30 g/cm^2 in Fig. 4.23 (dark vertical gray bar) may be related to some major turbidity currents perhaps associated with some other earthquakes. More information is required for further discussion on this point.

Recently, these lakes were investigated from the viewpoint of anthropogenic activities (Liu et al. 2014; Hillman et al. 2014; Zhang et al. 2010) and long-term climatic changes (Brenner et al. 1991; Hodell et al. 1999), although studies on tectonic movements (earthquake events) based on lacustrine sediments are limited for this area (Li et al. 1999). More interesting results will be given for environmental changes if research results are also interpreted with respect to the earth-surface

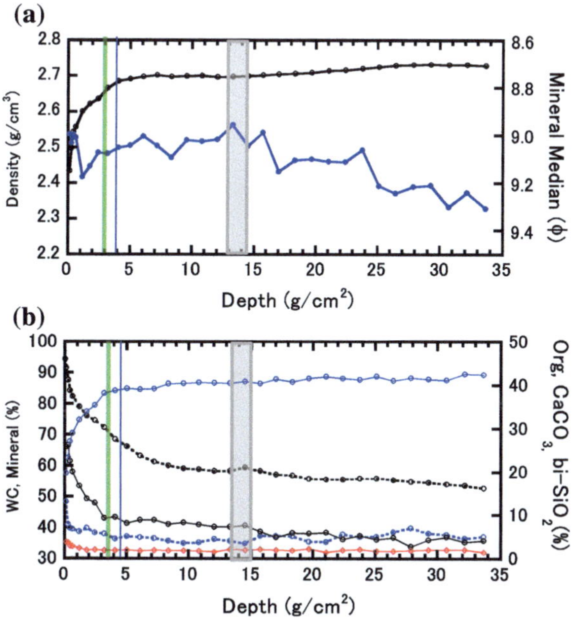

Fig. 4.21 Changes in analytical items for XY-4: **a** density; *black line*, and mineral median; *blue line*, and **b** water content (%); *dotted black line*, mineral content (%); *blue line*, organic content (%); *dotted blue line*, HCl-soluble content (%); *black line*, bi-SiO_2 content (%); *red line*. *Green vertical bar*; Tonghai earthquake, *blue vertical bar*; 1963, and *light gray vertical bar*; Songming earthquake

processes due to tectonic activity (historical and paleo-earthquakes, weathering, etc.) because this area is located in the tectonically active zone mentioned above.

The Kawauso-ike system in Kobe, discussed in the previous chapter (Chap. 3), was also influenced by the 1995 Kobe earthquake. Several short core samples were obtained before and after the earthquake. Two sediment traps were set on the pond floor just after the earthquake, and they were retrieved twice a month in summer and once in winter (Fig. 4.24). The observational results are going to be introduced later along with a model. Comparing core results with sediment trap results, the sedimentation rate was estimated (Kashiwaya et al. 2004; Fig. 4.25). The results show that the sedimentation rate rapidly increased just after the earthquake, and it gradually decreased afterward, although the grain size before and after the earthquake remained almost constant (Fig. 3.37a). There is little correlation between grain size and monthly rainfall (Fig. 4.26). To detect the earthquake effect on the sediments, three cores were obtained in 1999 and studied. The results are shown in Fig. 4.27, and they indicate that the grain size hardly changed around the earthquake time in the three cores, but the grain density slightly increased in all cores (Kashiwaya et al. 2015). Probably many fine mineral materials were produced in the catchment due to landslides caused by the earthquake. They were transported into the pond even during low rainfall periods, increasing the sedimentation rate and the grain density, although the grain size did not change.

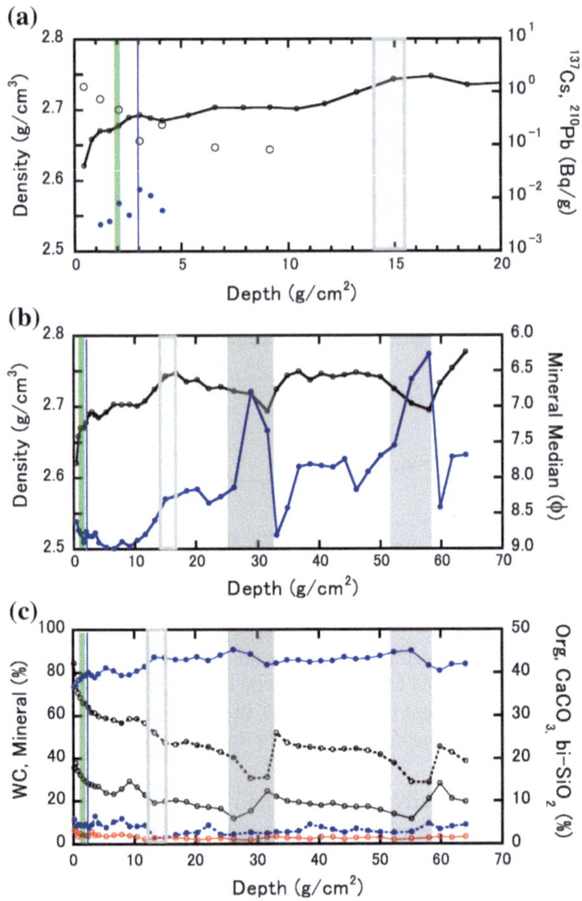

Fig. 4.22 Changes in analytical items for FX-1: **a** density; *black line*, ^{210}Pb concentration; *open circles*, and ^{137}Cs concentration; *blue circles*, **b** density; *black line*, and mineral median; *blue line*, and **c** water content (%); *dotted black line*, mineral content (%); *blue line*, organic content (%); *dotted blue line*, HCl-soluble content (%); *black line*, bi-SiO_2 content (%); *red line*. *Green vertical bar*; Tonghai earthquake, *blue vertical bar*; 1963, *light gray vertical bar*; Songming earthquake, and *dark gray vertical bar*; rapid mass movement

In general, most earth-surface processes related to tectonic activity, especially major ones, in catchments are also recorded in lacustrine sediments. Physical earth-surface changes due to the processes are linked to changes in not only the physical properties but also the chemical characteristics of the sediments in lake-catchment systems. However, distinguishing the impact of tectonic events from that of climatic events in sediments of the pre-instrumental observation period is difficult, especially in tectonically active and humid-climate areas such as Japan. It is important to acquire more precise information on earth-surface processes to address this issue further. This information may be obtained if materials including the information are found with suitable techniques and they are properly analyzed.

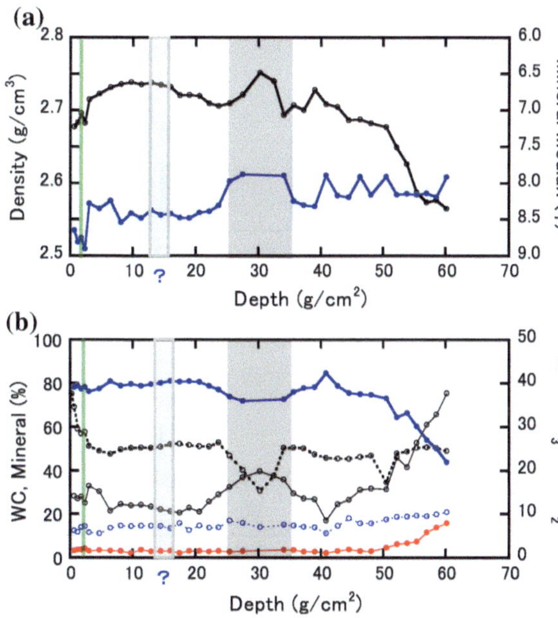

Fig. 4.23 Changes in analytical items for FX-2: **a** density; *black line*, and mineral median; *blue line*, **b** water content (%); *dotted black line*, mineral content (%); *blue line*, organic content (%); *dotted blue line*, HCl-soluble content (%); *black line*, and bi-SiO_2 content (%); *red line*. *Green vertical bar*; Tonghai earthquake, *light gray vertical bar*; Songming earthquake, and *dark gray vertical bar*; rapid mass movement

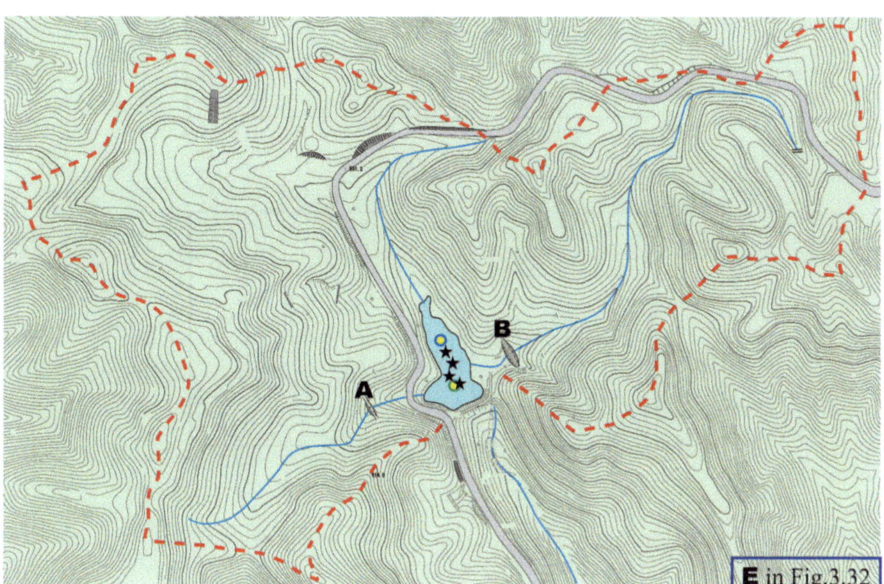

Fig. 4.24 Kawauso-ike lake-catchment system in Kobe, Japan. *Dotted red line*; catchment boundary, *A*, *B*; check dam, *stars*; sampling points and *circles*; sediment trap sites. *A* was constructed in 1984–1985 and *B* in 1989. Location E in Fig. 3.32

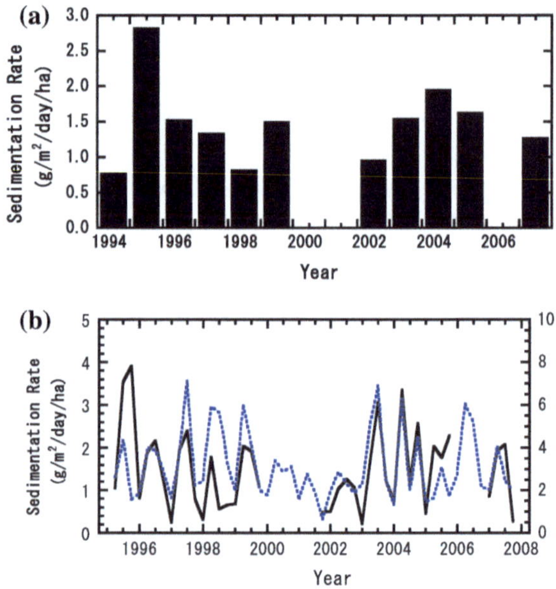

Fig. 4.25 Sedimentation rate with sediment traps; **a** yearly sedimentation rate, and **b** seasonal sedimentation rate (*solid black*) and rainfall (*dotted blue*)

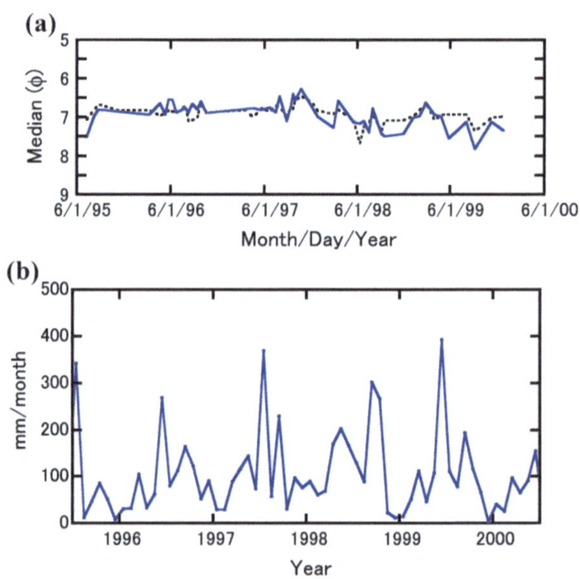

Fig. 4.26 **a** Temporal change in grain size (monthly and semimonthly) (*solid blue*; center site, and *dotted black*; back site), and **b** monthly precipitation

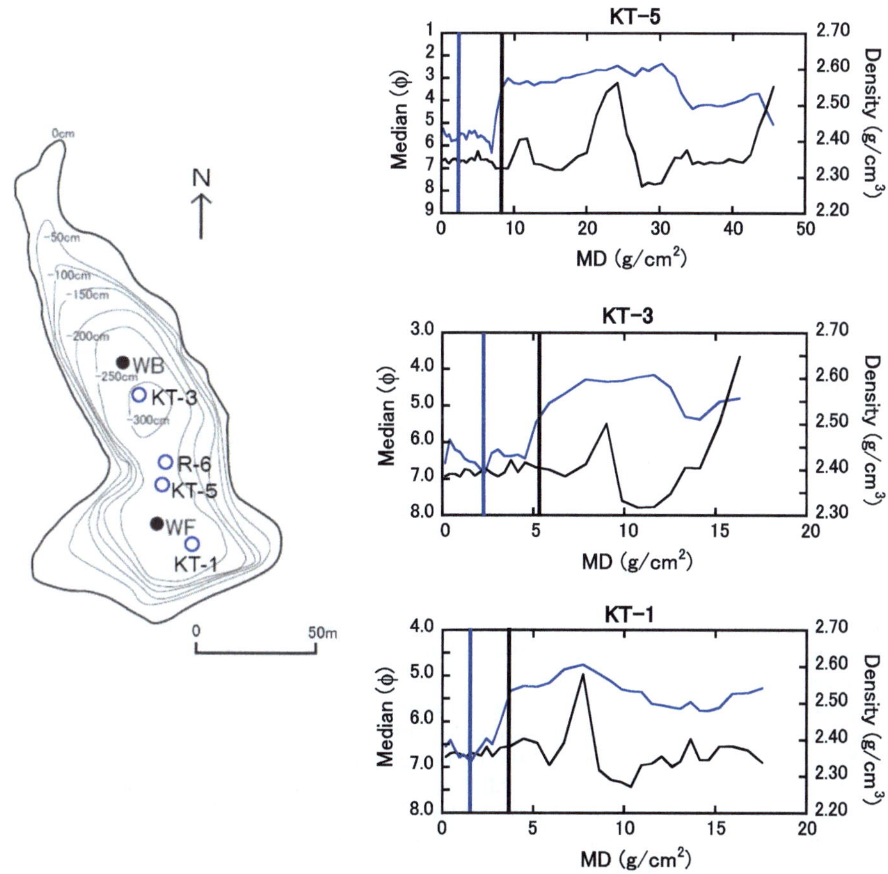

Fig. 4.27 Outline of Kawauso-ike (*left*). *Open circles*; sampling points and *closed circles*; sediment traps. *Blue line*; density (g/cm^3), *black line*; mineral median (ϕ), *black vertical bar*; 1984 (check dam construction), and *blue vertical bar*; 1995 (Kobe earthquake) (*right*)

References

The Asahi Shimbun (2001) Fuserareta chugoku daijisin (A hidden great earthquake in China) (Jan 30, 2001) (in Japanese)

BDP-99 Baikal Drilling Project Members (2005) A new Quaternary record of regional tectonic, sedimentation and paleoclimatic changes from drill core BDP-99 at Posolskaya Bank, Lake Baikal. Quat Int 136:105–121

Bezrukova E, Bukharov A, Bychinsky V, Colman S, Fedenya S, Gvozdkov A, Geletii V, Goreglyd A, Gorokhov I, Ivanov E, Kawai T, Kalmychkov G, Karabanov E, Kerber E, Khakhaev B, Khomutova M, Khursevich G, Kochukov V, Kravchinsky V, Krainov M, Krapivina S, Kudryashov N, Kuzmin M, Kulagina N, Letunova P, Levina O, Pevzner L, Prokopenko A, Scholz C, Solotchin P, Tkachenko O, Williams D (2004) High-resolution sedimentary record in a new BDP-99 core from Posol'sk Bank in Lake Baikal. Russ Geol Geophys 45(2):149–175

Brenner M, Dorsey K, Song X, Wang Z, Ruihua L, Binford MW, Whitmore TJ, Moore AM (1991) Paleolimnology of Qilu Hu, Yunnan Province, China. Hydrobiologia 214:333–340

FSGCR (Federal Service for Geodesy and Cartography of Russia) (2002) Map of Lake Baikal. Moscow, Russia (in Russian)

Hasebe N, Itono T, Katsuki K, Murakami T, Ochiai S, Katsuta N, Wang Y, Lee JY, Fukushi K, Ganzawa Y, Mitamura M, Tanaka K, Kim JY, Kashiwaya K (2015) Possible age models for Lake Onuma lacustrine sediments based on tuffs recovered in three cores. In: Kashiwaya K, Shen J, Kim JY (eds) Earth surface processes and environmental changes in East Asia—records from lake-catchment systems. Springer, Tokyo, 321p, pp 239–255

Hillman AL, Yu JQ, Abbott MB, Cooke CA, Baina DJ, Steinman BA (2014) Rapid environmental change during dynastic transitions in Yunnan Province, China. Quat Sci Rev 98:24–32

Hodell DA, Brenner M, Kanfoush SL, Curtis JH, Stoner JS, Song X, Wu Y, Whitmore TJ (1999) Paleoclimate of Southwestern China for the Past 50,000 yr inferred from lake sediment records. Quat Res 52:369–380

IKF (1996) The Baikal map. Irkutsk, Russia (in Russian)

Ishikawa K (2004) Climato-limnological changes inferred from core sediments of Lake Biwa, Japan. MS thesis for Kanazawa University (in Japanese with English abstract)

Itono T, Kashiwaya K, Ochiai S (2015) Reconstructing modern hydro-environmental fluctuations inferred from lacustrine sediment in Lake Onuma, Hokkaido. In: Kashiwaya K, Shen J, Kim JY (eds) Earth surface processes and environmental changes in East Asia—records from lake-catchment systems. Springer, Tokyo, 321p, pp 239–255

Iwamura T, Inoue K, Nishiyama A (2002) Biwako-seigan earthquake (1662) and Machii landslide dam. Hist Earthq 18:52–58 (in Japanese with English abstract)

Japan Meteorological Agency, Volcanological Society of Japan (eds) (2013) National catalogue of the active volcanoes in Japan (the fourth edition, English version). Japan Meteorological Agency, Tokyo

Kashiwaya K (2012) Earth surface processes and environmental changes in lake-catchment systems. Trans Jpn Geomorphol Union 33:121–136

Kashiwaya K, Kusumoto T, Tang C (2006) Hydro-geomorphological fluctuations and changes in population and land use. In: Sakamoto M, Kumagai M (eds) Lakes and their catchments in East Asian monsoonal zone. Nagoya University Press, Nagoya, 347p, pp 234–258 (in Japanese)

Kashiwaya K, Tsuya T, Okimura T (2004) Earthquake-related geomorphic environment and pond sediment information. Earth Surf Proc Land 29:785–7931

Kashiwaya K, Okimura T, Itono T, Ishikawa K, Kusumoto T (2015) Present earth-surface processes and historical environmental changes inferred from lake-catchment systems. In: Kashiwaya K, Shen J, Kim JY (eds) Earth surface processes and environmental changes in East Asia—records from lake-catchment systems. Springer, Tokyo, 321p, pp 1–24

Komatsubara T (2006) 1662 Kanbun Ohmi-Wakasa Earthquake and histo-geographical investigation of disaster areas. Historical Disaster Studies in Kyoto No. 5, pp 31–38 (in Japanese)

Komatsubara T (2012) Re-examination of source faults of the Genryaku Ohmi-Yamashiro Earthquake in 1185. Hist Earthq 27:1–7 (in Japanese with English abstract)

Kusumoto T (2003) Hydro-geomorphological environment in the eastern margin of the Tibetan Plateau inferred from geomorphic information and lake sediment information. MS thesis for Kanazawa University

Li JS, Song XL, Sun YL, Zhang ZX, Song YD, Liu G (1999) The magnetic susceptibility measurements of turbidity current sediments from Fuxian Lake of Yunnan Province and their correlations with earthquakes. Acta Seismol Sin 12:93–98

Liu W, Wu J, Zeng H, Ma L (2014) Geochemical evidence of human impacts on deep Lake Fuxian, southwest China. Limnologica 45:1–6

Mats VD, Perepelova TI (2011) A new perspective on evolution of the Baikal Rift. Geosci Front 2:349–365

Matsui J (1988) Radiocarbon age of peat at the Dairaike in Omi Imazu, Shiga Prefecture. Quat Res 26:407–408 (in Japanese with English abstract)

Meyers PA, Takemura K, Horie S (1993) Reinterpretation of late quaternary sediment chronology of Lake Biwa, Japan, from correlation with marine glacial-interglacial cycles. Quat Res 39:154–162

Mizuyama T, Ikeda H, Ohashi T (1975) Ohmi bonchi to Biwako shuhen no chikei (Geomorphology in the Ohmi basin and Lake Biwa). Construction Ministry Kinki Regional Construction Bureau, Otsu 267p (in Japanese)

Murai I (1962) A brief note on the eruption of the Yake-dake volcano of June 17, 1962. Bull Earthq Res Inst Univ Tokyo 40:805–814

Nishiyama A (2006) Kanbun 2 nen (1662) Ohmi-Wakasa jishin niokeru kyoto deno higai to shinsai taio (The 1662 Kanbun Ohmi-Wakasa earthquake, and damage by the earthquake and measures to it in Kyoto). Historical Disaster Studies in Kyoto No. 5, 39–54 (in Japanese)

Nishiyama A, Komatsubara T (2006) The damage in the Kyoto-basin due to the 1662 Kanbun Ohmi-Wakasa earthquake. Hist Earthq 21:165–171 (in Japanese with English abstract)

Ochiai S, Nagao S, Itono T, Suzuki T, Kashiwaya K, Yonebayashi K, Okazaki M, Kaeriyama M, Qin YX, Hasegawa T, Yamamoto M (2015) Recent eutrophication and environmental changes in the catchment inferred from geochemical properties of Lake Onuma sediments in Japan. In: Kashiwaya K, Shen J, Kim JY (eds) Earth surface processes and environmental changes in East Asia—records from lake-catchment systems. Springer, Tokyo, 321p, pp 257–268

Oikawa T, Okuno M, Mitsuru Okuno Nakamura T (2002) The past 3000 years' eruption history of Yakedake volcano in the Northern Japan Alps. J Geol Soc Jpn 108:88–102

Okuda S, Suwa H, Okunishi K, Yokoyama K, Nakano M (1980) Observation of the motion of debris flow and its geomorphological effects. Zeitschrift fur Geomorphology. Suppl Bd 35: 142–163

Shimada T, Kashiwaya K, Masuzawa T, Hyodo M (2002) Hydro-environmental fluctuation in a fake—catchment system during the late Holocene inferred from Lake Yogo sediments. Trans Jpn Geomorphol Union 23:415–431 (in Japanese with English abstract)

Sun W, Shen J, Zhang E, Hasebe N, Kashiwaya K, Chen R, Itono T (2016) Stable nitrogen isotope record of lacustrine sediments in Lake Onuma (Northern Japan) indicates regional hydrological variability during the past four centuries. Quat Int 397:307–316

Suwa H (1988) Focusing mechanism of large boulders to a debris-flow front. Trans Jpn Geomorphol Union 9:151–178

Takamatsu T, Takada J, Tanaka A, Kawai T (2003) Environmental changes during the past 5 million years inferred from elemental composition of Baikal sediments (BDP93 and BDP96). Chikyu Monthly 42:108–118 (in Japanese)

Usami T (2003) Saishinban Nihon higai-jishin soran (An overview of destructive earthquakes in Japan—the latest edition). University of Tokyo Press, Tokyo, 605p (in Japanese)

Wan SM, Dou HS (1998) Lake records of China. Science Press, Beijing, 580p (in Chinese)

Yang ZX, Waldhauser F, Chen YT, Richards PG (2005) Double-difference relocation of earthquakes in central-western China, 1992–1999. J Seismolog 9:241–264

Yokota Y (2011) Shiozu-ko iseki ni miru Biwako no suiihendo (Water level fluctuation in Lake Biwa inferred from the ruins of ancient Shiozu Port). In: Hayashi Hiromichi Sensei Tainin-kinen-ronshu Kanko-kai (Publication board for a festschrift of Prof. H. Hayashi retirement) (ed) Biwako to chiiki-bunka (Lake Biwa and local culture). Sun Rize Press, Hikone, 458p, pp 312–317 (in Japanese)

Yokoyama T (1984) Stratigraphy of the quaternary system around Lake Biwa and geohistory of the ancient Lake Biwa. In: Horie S (ed) Lake Biwa. Dr. W. Junk, The Hague, 654p, pp 43–128

Yoshimoto M, Ui T (1998) The 1640 sector collapse of Hokkaido-Komagatake volcano, northern Japan. Bull Volcanol Soc Jpn 43:137–148 (in Japanese with English abstract)

Zhang H, Li S, Feng Q, Zhang S (2010) Environmental change and human activities during the 20th century reconstructed from the sediment of Xingyun Lake, Yunnan Province, China. Quat Int 212:14–20

Chapter 5
Anthropogenic Forces on Systems

Probably, the first impact of human activities on lake-catchment systems was related to irrigation. Reservoirs for irrigation may have been among the first agricultural constructions. For example, irrigation with reservoirs was developed in Ancient Egypt and Mesopotamia (Shanan 1987). Such reservoirs were constructed to establish new lake-catchment systems, and sediments accumulated there may include precious information on historical environmental changes in the systems since their construction. However, long continuous sedimentary records are rare because of physical destruction due to natural and artificial activities at the inflow and outflow sites. Generally, continuous records of such systems are limited to certain intervals, as described below.

The Pyeokgolje reservoir in Kimje, southern Korea, was established in 330 AD along with a rich rice field downstream. Traditionally, the construction of this reservoir was said to have made the Honam Plain (the largest plain in Korea) fertile (tasty rice-producing area; a famous Korean dish "Bibimbap" was born here). Recently, however, Park et al. (2003) proposed that this reservoir was constructed to prevent tidal waves from reaching the inland area. The Euirimji reservoir in Jecheon, central Korea, was also established for irrigation initially, at approximately 2000 years ago and it was constructed anew approximately 1200 years ago; some environmental changes were reconstructed with sedimentary records for the past 2000 years (Kim et al. 2012, 2015; Nahm et al. 2012).

The impacts of anthropogenic activities on such systems are observed in numerous areas around the world although modern dam construction itself has a significant influence on the regional environment. The construction of the Aswan High Dam in 1964 has significantly affected not only on the Nile catchment but also

K. Kashiwaya, *Geomorphology of Lake-Catchment Systems*,
Environmental Earth Sciences, DOI 10.1007/978-981-10-5110-4_5

on the Mediterranean Sea (Nixon 2004; Stanley 1996). The Three Gorges Dam in the Chang jiang, China, was completed in 2009, but the environmental repercussions (e.g., on the local climate and ecosystem) from its construction remain (Wu et al. 2006). The new lake-catchment system accumulates much information on environmental problems occurring in this system. The problems of the Aral Sea related to recent human activities have also attracted the keen attention of scientists as well as policy-makers. The rapid decrease in the water area of the Aral Sea since ca 1960 was connected to the overuse of water for irrigation in the catchment (along the Amu Darya and the Syr Darya) (Micklin 2010). Moreover, increase and decrease in the water area of the Aral Sea has repeatedly occurred during the Holocene (Fig. 5.1; Boomer et al. 2000) suggesting that paleoenvironmental data may be also used for understanding the recent artificial land use changes in this area.

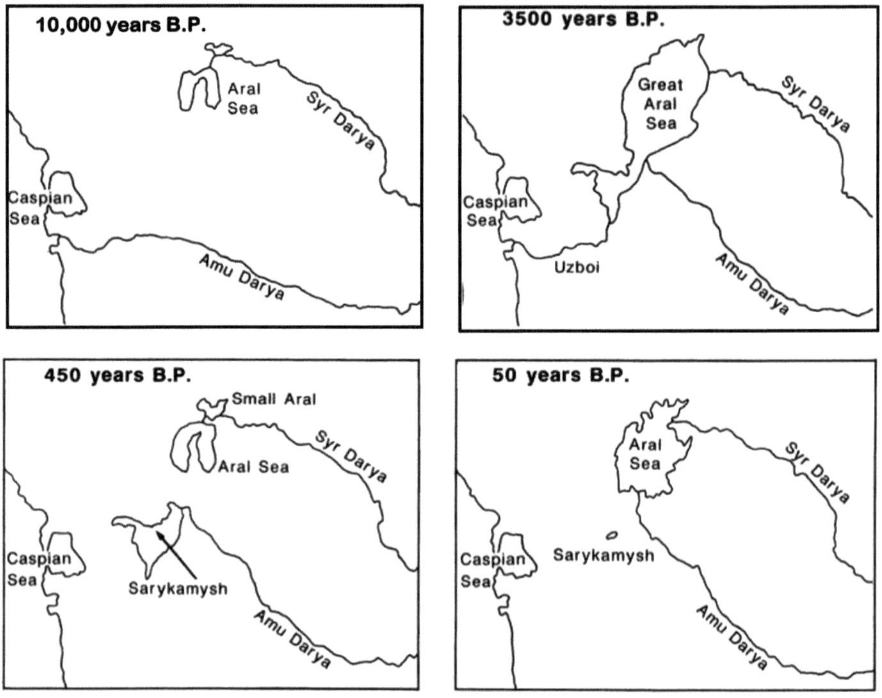

Fig. 5.1 Geographical changes in the Aral Sea during the past 10,000 years (modified after Boomer et al. 2000)

Fig. 5.2 Lake Riyuetan (Sun Moon Lake; *circle*) in Taiwan (*left*). Circle means sampling point (*right*) (after Ochiai et al. 2012)

One of the biggest construction works for a lake-catchment system in the world (the first biggest construction in Asia) at the early twentieth century was the Riyuetan Power Plant station in Taiwan. Lake Riyuetan (Sun Moon Lake) was dammed for electrical power in 1934 with additional inflow from a different catchment (the Zhuoshui; the largest river in Taiwan) (Fig. 5.2). Most inflow has been artificially controlled since then, and a new lake-catchment system was established. Changes in the hydrological conditions due to the dam construction are clearly detected in the physical properties of the lacustrine sediments, which were sampled in 2012 (Fig. 5.3). In particular, rapid increase in grain density and decrease in grain size were observed after the dam construction at a depth of 25–30 cm for both long and short cores, as well as in a core obtained in 1995 (Lu et al. 2009).

Themes on lake (reservoir)-catchment systems with lacustrine sediments were earnestly studied in the 1980 s and 1990 s (e.g., McManus and Duck 1993), including some artificial forces although quantitative studies for future prediction were limited. A few of the above-mentioned studies conducted in Japan also yielded some information on artificial forces. An artificial flume was constructed

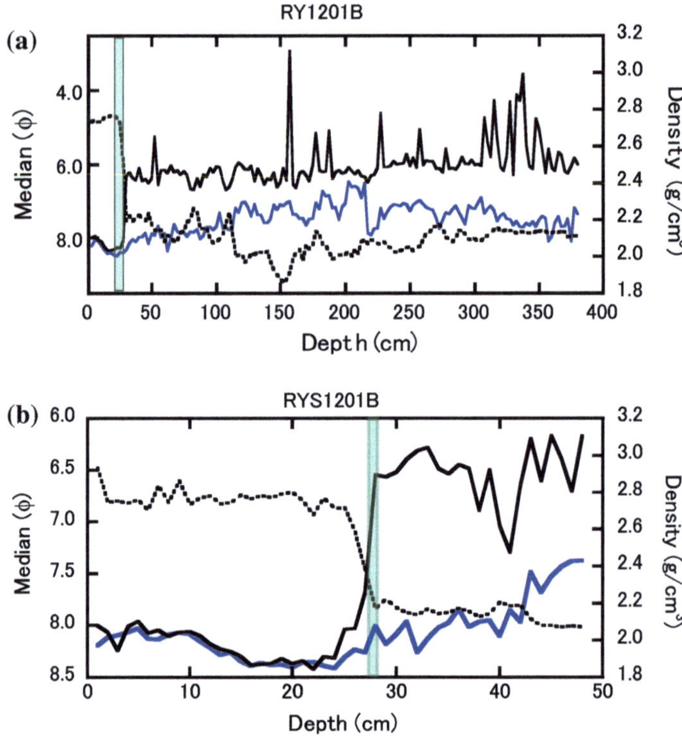

Fig. 5.3 **a** Changes in physical properties in a long core (RY1201B) and **b** a short core (RYS1201B). *Solid black line*, sediment grain size (median); *blue line*, mineral grain size (median); *dotted black line*, density; and *green bar*, dam construction time; (original data after Liu 2013)

into Lake Yogo to acquire water from the river Yogo for irrigation and flood control in 1960. Then, the catchment area increased by approximately four times before the construction (an enlarged lake-catchment system was established). This anthropogenic activity was clearly detected in the rapid change of the sedimentation rate in 1960, as shown in Fig. 5.4 (Shimada et al. 2002). This also suggests that discharge increased (sedimentation rate increased) due to the enlargement of the catchment area, whereas the grain size apparently corresponded to precipitation after the flume construction; probably, fine materials were mainly transported into the lake, under similar sediment transport conditions as those before the construction because of sediment control in the flume (Fig. 5.5).

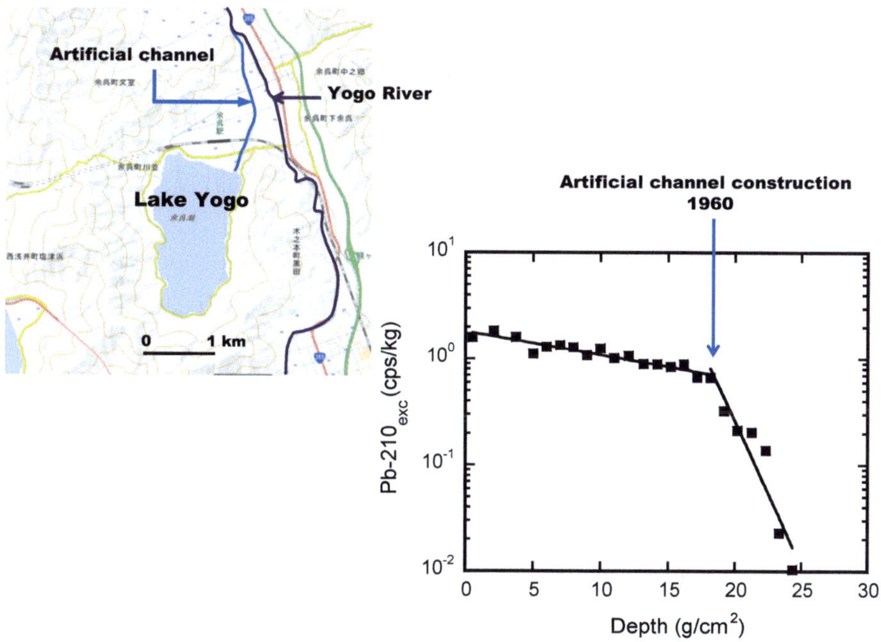

Fig. 5.4 Artificial channel construction in 1960 (*left*) (based on GSI MAP) and change in sedimentation rate (*right*) (modified after Shimada et al. 2002)

Anthropogenic activities also affected the lacustrine sediments of the Kawauso-ike system in Kobe, as mentioned above (Fig. 4.27). A rapid decrease in grain density was detected for the two cores obtained at before and after the 1995 earthquake, at a depth of approximately 2.3–2.5 g/cm^2 from the bottom surface for the core obtained in 1986, and at 8–9 g/cm^2 for the core obtained in 1999 (black lines in Fig. 5.6; Kashiwaya et al. 2015). Parts of rapid density decrease (black lines) for the three cores obtained 1999 (Fig. 4.27) perhaps corresponded to 1984–1985, when the check dams were constructed (Fig. 4.24). Materials in the west tributary flowed into Kawauso-ike directly suggesting that the constructed check dam could play an important role in trapping the materials produced in the catchment. In contrast, materials in the east tributary flowed into the pond via a delta formed at the mouth of the river, where some materials might have been used for developing the delta. The potential volume of the dam deposition for the west tributary was 540 m^3, and for the east tributary, it was 5561 m^3 when the check dams were constructed (Kobe City Government). Generally, check dams prevent

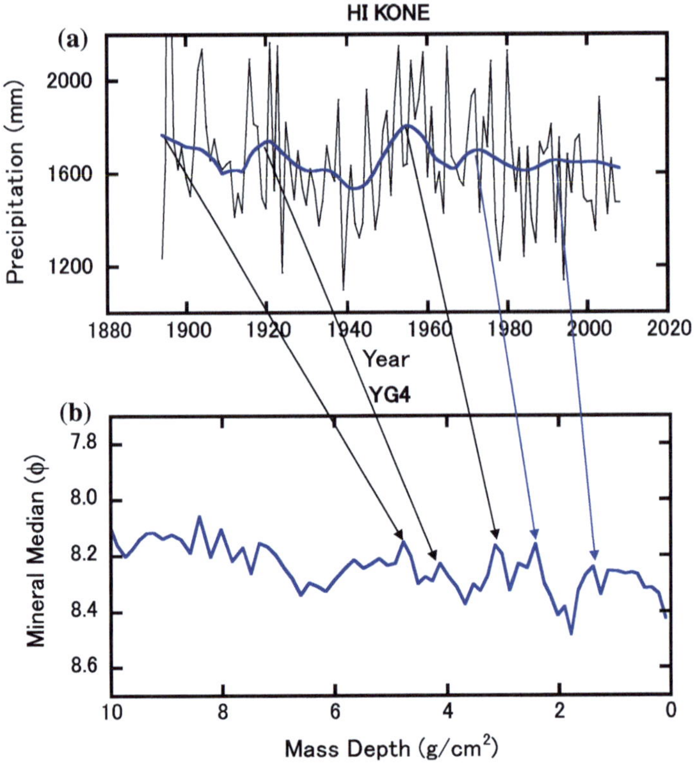

Fig. 5.5 **a** Annual precipitation (*thin line*, original; and *blue thick line*, filtered) in Hikone Regional Observatory and **b** change in the grain size of Lake Yogo sediments (mineral median) (modified after Shimada et al. 2002)

materials, especially large ones, from flowing into reservoirs. This may have been also related to a limited grain size increase after the 1995 earthquake although the volume of the mineral material increased.

Land transformation in systems also significantly influences the catchment conditions and sedimentation. Recently, several large land transformation schemes have been implemented, especially in economically growing areas in the world. There were also large land transformations in Japan from the 1960s to 1980s. Large transformation areas in the Kobe district were developed for artificial islands (port

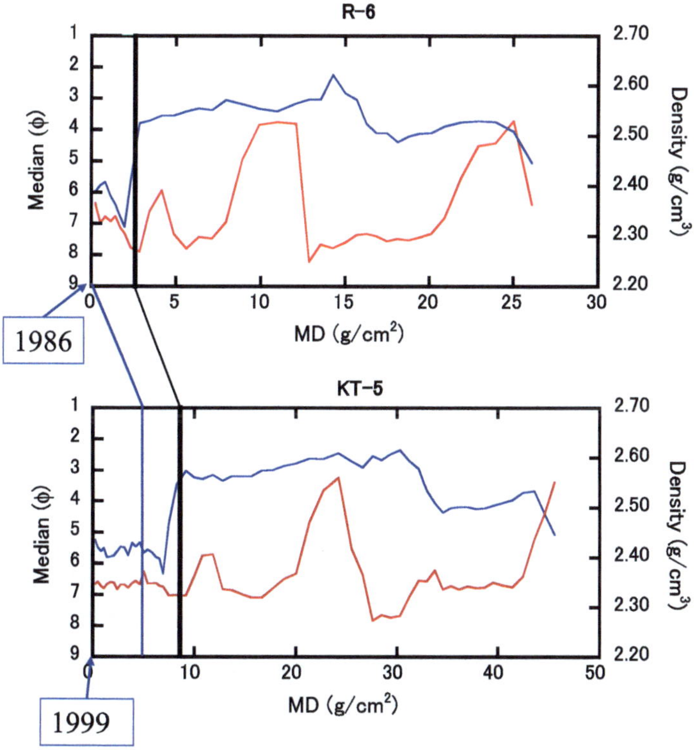

Fig. 5.6 Changes in grain size (median) and density for the two cores obtained before Kobe earthquake (1986; R-6) and after the earthquake (1999; KT-5) (modified after Kashiwaya et al. 2015)

and airport) and residence areas. A lake-catchment system (Kawa-ike system with two subsystems) was selected to investigate the changes in the physical environment with land transformation (Fig. 5.7; Kashiwaya et al. 1997). This lake-catchment system was observed for five years. Materials were artificially eroded and transported for creating a flat area and artificial islands near Kobe. Temporal changes in land transformation (increase in naked area at the early stage of the transformation and gradual decrease with increasing artificial vegetation cover in the late stage) were observed, and the physical properties of sediments obtained with sediment traps were measured. The sedimentation rate in the system with hard land transformation is much larger than the rate without land

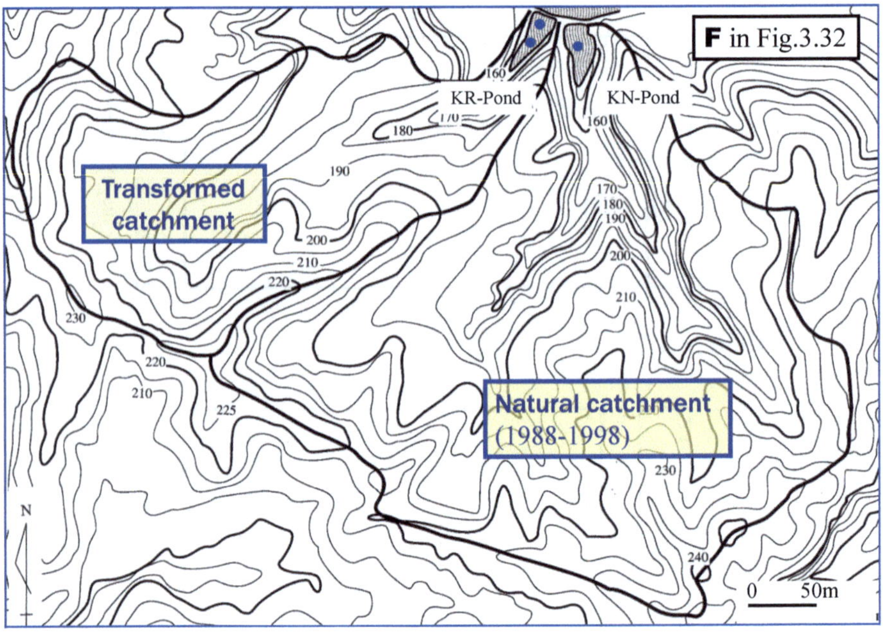

Fig. 5.7 Outline of Kawa-ike lake-catchment system (original map by the Development Bureau of the Kobe City Government). Location F in Fig. 3.32

transformation is (Fig. 5.8a), and it corresponds to changes in the development stage (Fig. 5.8b). The sedimentation rate pattern is similar in the two ponds because the lowest ends of the ponds are connected. Grain density increases in the early stage of development and decreases in the late stage, whereas the loss on ignition slightly decreases in the early stage and increases afterward (Fig. 5.9). An increase in grain density is closely related to mineral material production due to the land transformation, and it is a good indicator of the development stage (catchment condition). A change in the loss on ignition is probably related to vegetation cover (natural vegetation and artificial vegetation).

As mentioned in the above chapter, the Onuma lake-catchment system offers valuable information on the tectonic activity and the climate of this area. This system also includes recent information on anthropogenic activities. Several short cores in Lake Onuma also show some artificial changes in the catchment; increase

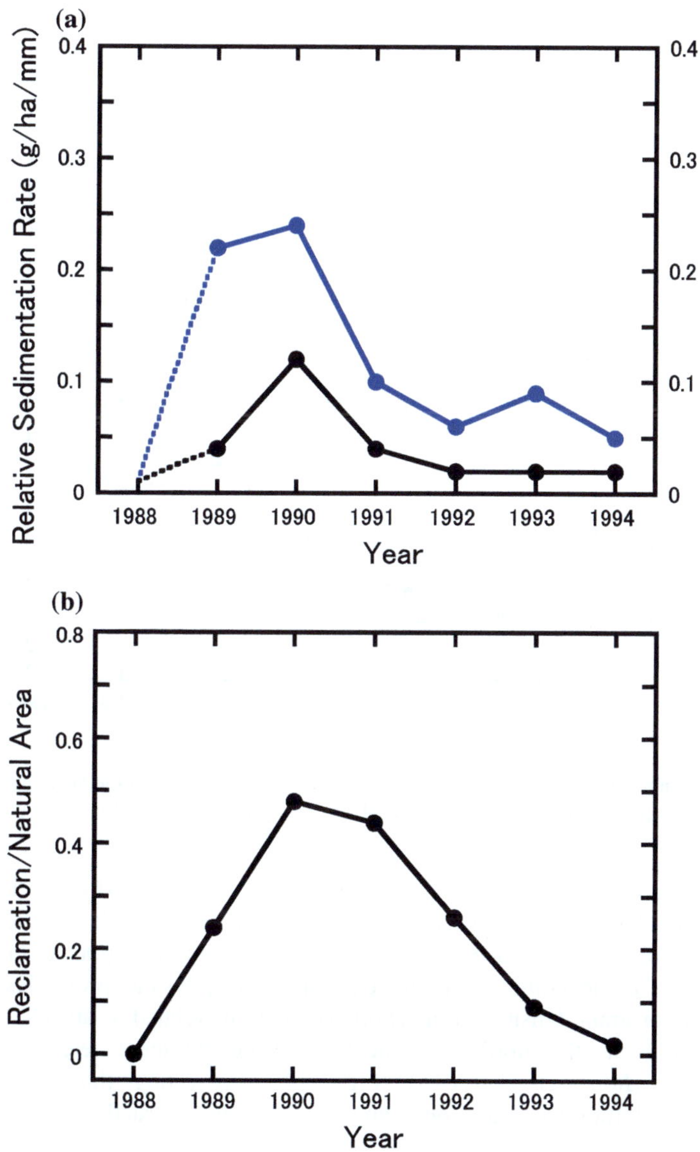

Fig. 5.8 **a** Relative sedimentation rate in tow ponds (KR pond and KN pond) (*blue line*, reclamation area (KR pond); and *black line*, natural area (KN pond)); and **b** ratio of reclamation area to vegetation area. Reproduced from Kashiwaya et al. (1997) by permission of John Wiley & Sons Ltd

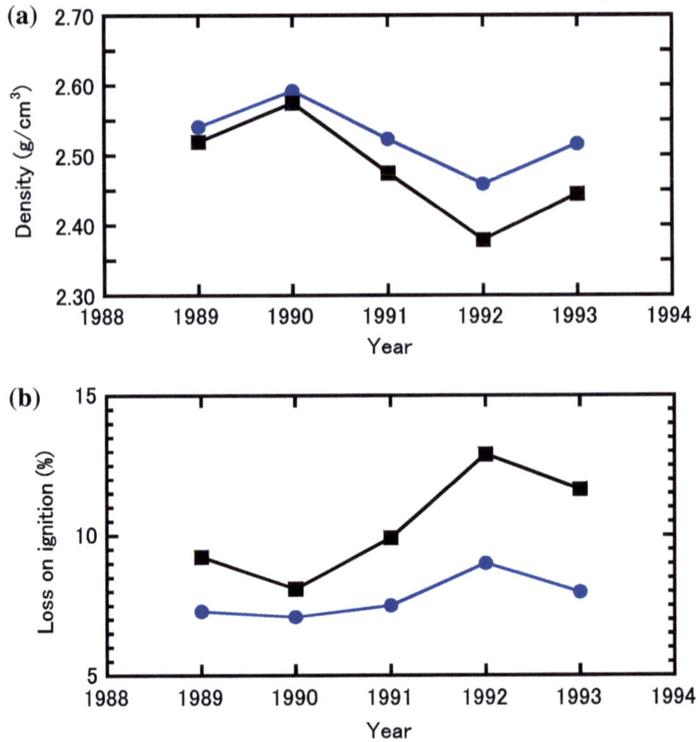

Fig. 5.9 Changes in **a** density and **b** loss on ignition. *Blue line*, reclamation area; and *black line*, natural area. Reproduced from Kashiwaya et al. (1997) by permission of John Wiley & Sons Ltd

in the $\delta^{15}N$ concentration in the recent period may be attributed to the increase in the livestock farming activities in the south catchment, considering the relations among C/N ratio, $\delta^{15}N$ concentration, and $\delta^{13}C$ concentration (Ochiai et al. 2015) (Fig. 5.10). On the contrary, an increase in some physical properties (mineral content and/or grain density) in the short core may be related to artificial landform transformation in the north catchment (land development, etc.) as well as precipitation (Itono et al. 2015) (Fig. 5.11), suggesting that anthropogenic activities in addition to natural phenomena have been recorded in these lacustrine sediments.

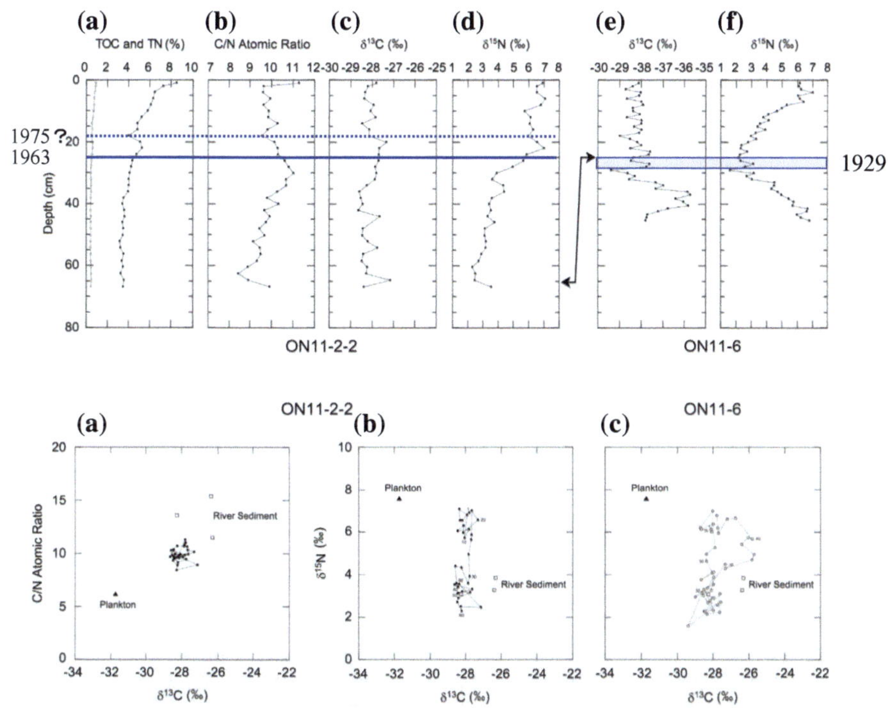

Fig. 5.10 Changes in chemical properties of lacustrine sediments from Lake Onuma: the *upper* figures—**a** TOC and TN contents, **b** C/N atomic ratio, **c** carbon isotope ratio $\delta^{13}C$, **d** nitrogen isotope ratio $\delta^{15}N$ for the ON11-2-2 core, **e** carbon isotope ratio $\delta^{13}C$, and **f** nitrogen isotope ratio $\delta^{15}N$ for the ON11-6 core. Scatter plots in the *lower* figures: **a** $\delta^{13}C$–C/N ratio, **b** $\delta^{13}C$–$\delta^{15}N$ plots of ON11-2-2 sediment, planktonic material, and river sediment, and **c** the $\delta^{13}C$–$\delta^{15}N$ plots of ON11-6 sediment. *Circles* indicate lacustrine sediment samples. *Triangle* and *square* indicate planktonic material and river sediment, respectively (modified after Ochiai et al. 2015)

Most large anthropogenic activities (forces) affecting the systems at present may be easily detected in lacustrine sediments. However, it is more difficult to isolate their effects on past lake-catchment systems, partly because such activities were limited and relatively small scale, compared with recent large activities. Nevertheless, information on present activities is of great use for understanding past natural processes as well as anthropogenic processes.

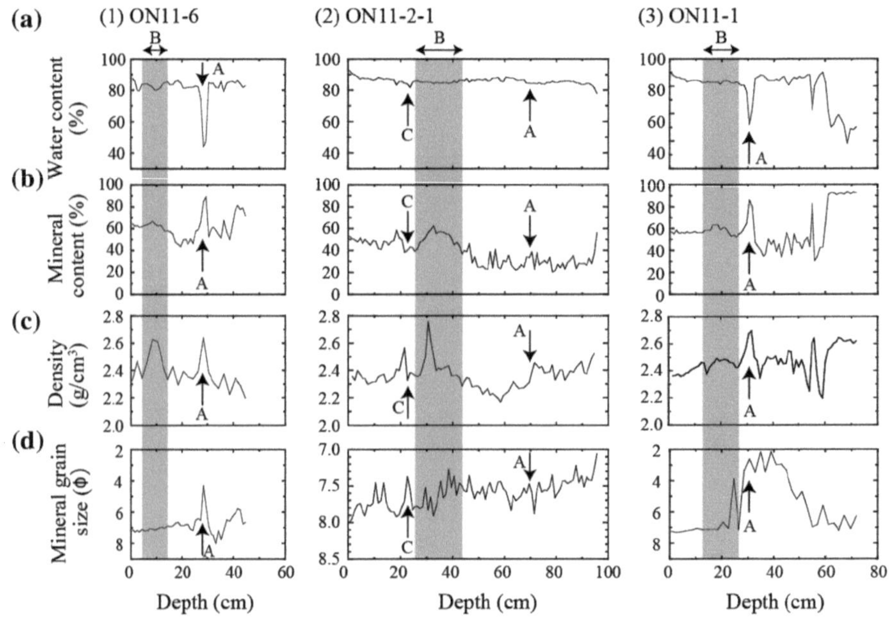

Fig. 5.11 Physical properties of short-core sediments from Lake Onuma. (*1*) ON11-6, (*2*) ON11-2-1, and (*3*) ON11-1. **a** Water content (%), **b** mineral content (%), **c** density (g/cm^3), and **d** mineral grain size (ϕ). *Arrow-A* shows the estimated eruption of Mt. Komagatake in 1929, *belt-B* shows the layer of peak in mineral content and density, and *arrow-C* shows the estimated typhoon in 1975 (modified after Itono et al. 2015)

References

Boomer I, Aladin NV, Plotnikov I, Whatley R (2000) The palaeolimnology of the Aral Sea: a review. Quat Sci Rev 19:1259–1278

Itono T, Kashiwaya K, Ochiai S (2015) Reconstructing modern hydro-environmental fluctuations inferred from lacustrine sediment in Lake Onuma, Hokkaido. In: Kashiwaya K, Shen J, Kim JY (eds) Earth surface processes and environmental changes in East Asia—records from lake-catchment systems. Springer, Tokyo, 321p, pp 239–255

Kashiwaya K, Okimura T, Harada T (1997) Land transformation and pond sediment information. Earth Surf Proc Land 22:913–922

Kashiwaya K, Okimura T, Itono T, Ishikawa K, Kusumoto T (2015) Present earth-surface processes and historical environmental changes inferred from lake-catchment systems. In: Kashiwaya K, Shen J, Kim JY (eds) Earth surface processes and environmental changes in East Asia—records from lake-catchment systems. Springer, Tokyo, 321p, pp 1–24

Kim JY, Nahm WH, Yang DY, Shin JH, Lee JY, Hong SS, Kim JK, Lee HJ, Krivonogov S, Safonova I (2012) The Eurim Lake and its hosting valley: a unique ancient reservoir with bank constructions in Korea. Trans Jpn Geomorphol Union 33:137–148

Kim JY, NahmWH, Yang DY, Hong SS, Yi SH, Choi HW, Lee JY, Kim JK, Kim JC, Jo KN, Katsuki K, Park HS, Kashiwaya K, Hasebe H, Fukushi K, Endo N, Shen J, Wang Y, Oh KC (2015) Paleohydrological fluctuation of the historical Eurimji reservoir lake. In: Kashiwaya K, Shen J, Kim JY (eds) Earth surface processes and environmental changes in East Asia—records from lake-catchment systems. Springer, Tokyo, 321p, pp 143–161

Liu SH (2013) Sediment characteristic analysis and discussion of environmental changes since 16 ka—a case study at Sun-Moon Lake, Taiwan. Master's thesis, National Taiwan University, 132p (in Chinese with English abstract)

Lu W, Liew P, Li H (2009) Paleoenvironmental study of the Yueh Tan Core II in Sun Moon Lake, central Taiwan. West Pac Earth Sci 9:121–138

McManus J, Duck RW (eds) (1993) Geomorphology and sedimentology of lakes and reservoirs. Wiley, Chichester, 278p

Micklin P (2010) The past, present, and future Aral Sea: Lakes & Reservoirs. Res Manag 15:193–213

Nahm WK, Kim JK, Kim JY, Lim J, Kim JC, Yu KM (2012) Topographical evolution and ^{14}C age dating of the construction of the Eurimji reservoir (Jecheon, Korea). J Archaeol Sci 39:3706–3713

Nixon SW (2004) The artificial Nile: the Aswan High Dam blocked and diverted nutrients and destroyed a Mediterranean fishery, but human activities may have revived it. Am Sci 92 (2):158–165

Ochiai S, Lin JC, Kashiwaya K, Jen CH (2012) Influence of construction of dams and waterway tunnels on sedimentation rate and bottom topography in Sun Moon Lake, Taiwan. Trans Jpn Geomorphol Union 33:149–170

Ochiai S, Nagao S, Itono T, Suzuki T, Kashiwaya K, Yonebayashi K, Okazaki M, Kaeriyama M, Qin YX, Hasegawa T, Yamamoto M (2015) Recent eutrophication and environmental changes in the catchment inferred from geochemical properties of Lake Onuma sediments in Japan. In: Kashiwaya K, Shen J, Kim JY (eds) Earth surface processes and environmental changes in East Asia—records from lake-catchment systems, 321p, pp 257–268

Park SH, Choi KW, Lee KY, Um MC, An JS (2003) A study on possibility of the Byeokgolje Dam as a sea dike. KCID J 10(1):64–72 (in Korean with English abstract)

Shanan L (1987) The impact of irrigation. In: Wolman MG, Fournier FGA (eds) Land transformation in agriculture. Wiley, Chichester, 531p, pp 115–131

Shimada T, Kashiwaya K, Hyodo M, Masuzawa T (2002) Hydro-environmental fluctuation in a lake-catchment system during the late Holocene inferred from Lake Yogo sediments. Trans Jpn Geomorphol Union 23:415–431

Stanley DJ (1996) Nile delta: extreme case of sediment entrapment on a delta plain and consequent coastal land loss. Mar Geol 129:189–195

Wu L, Zhang Q, Jiang Z (2006) Three Gorges dam affects regional precipitation. Geophys Res Lett 33:L13806

Chapter 6
Observations on Lake-Catchment Systems and Experimental Models

6.1 Experimental Models Based on Field Observations of Lake-Catchment Systems in the Reclamation Area

Firstly, we introduce the relations between the observed items. Quantitative expressions among them are required to derive the causes and effects. These are called experimental (empirical) models (equations). Some models are used as first approximate ones for establishing next causal ones. Others are used, as they are, for empirical predictions in the near future. For example, many experimental models have been proposed on slope erosion, such as the Universal Soil Loss Equation (USLE, RUSLE) (e.g., Jones et al. 1996). They are often used for the discussion and prediction on present erosion problems. They are empirically available for some slope systems. This is because erosion processes in the systems are complex, and it is difficult to express the processes with simple physical models.

An experimental equation for the sedimentation rate is proposed for the Kawa-ike systems in Kobe based on the field data, pertaining to the above-mentioned landform transformations (Fig. 5.7; Kashiwaya et al. 1997). In this model, it is assumed that the sedimentation in ponds is mainly related to three factors: the catchment conditions, under which the sediment materials are produced (the catchment area conditions: slope, valley density, vegetation cover, etc.); the erosional conditions (rainfall intensity); and the internal pond conditions (water level, bed morphology, etc.). The relation is roughly expressed as

$$SD(t) = L_f(t) \times D_f(t) \times C_f(t) \tag{6.1}$$

where $SD(t)$ is the sedimentation rate, $L_f(t)$, the lake (pond) factor, $D_f(t)$, the drainage (catchment) factor, and $C_f(t)$ the climatic (rainfall) factor at time t. When changes in the pond conditions are limited for a certain interval (several years), Eq. (6.1) may be assumed

K. Kashiwaya, *Geomorphology of Lake-Catchment Systems*,
Environmental Earth Sciences, DOI 10.1007/978-981-10-5110-4_6

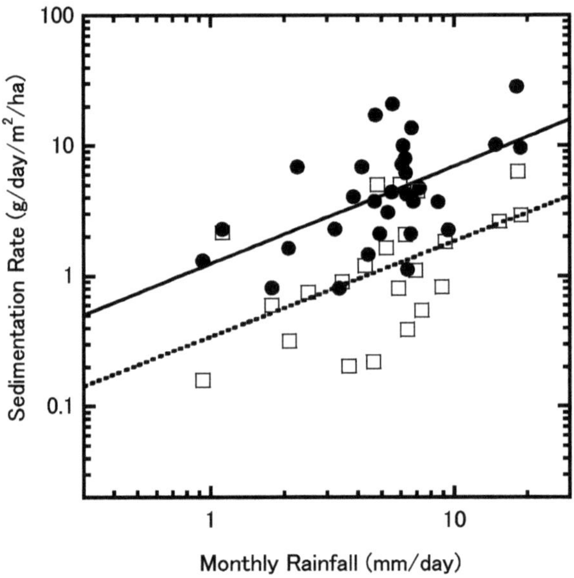

Fig. 6.1 Relationship between monthly sedimentation rate (g/day/m^2/ ha) and monthly rainfall (mm/day). *Closed circles* and *solid line*; reclamation area (KR), and *squares* and *dotted line*; natural area (KN). Regression equations for KR; $Y = 1.24X^{0.748}$ ($R = 0.53$) and for KN; $Y = 0.34X^{0.731}$ ($R = 0.52$), Y; sedimentation rate and X; monthly rainfall

$$SD(t) \propto D_f(t) \times C_f(t). \tag{6.2}$$

The validity of this relation is checked with the monthly sedimentation rate (per unit catchment area for canceling catchment size effect; g/day/m^2/ha) and monthly rainfall (mm/day). Snowfall in winter may be neglected in this area. Results are shown in Fig. 6.1, indicating that the monthly sedimentation rate (KR and KN) is roughly proportional to the monthly rainfall, and KR is more sensitive to rainfall intensity than KN, suggesting that the catchment factor in KR is larger (more erodible) than KN. If the pond factor ($L_f(t)$) is assumed constant ($=L_f$), (6.1) is modified to

$$D_f(t) = SD(t)/\left[L_f \bullet C_f(t)\right]. \tag{6.3}$$

This shows that the erodibility of surface catchment materials (system movable factor) is expressed as a relative catchment factor, shown as

$$RD_f(t) = SD(t)/C_f(t). \tag{6.4}$$

Average yearly values of $RD_f(t)$ are shown in Fig. 6.2. The values for the KR sediment are much larger than those for KN. This means that materials were more easily transported to the KR pond from the KR catchment than from the KN catchment. It also indicates that the relative catchment factor is closely related to the

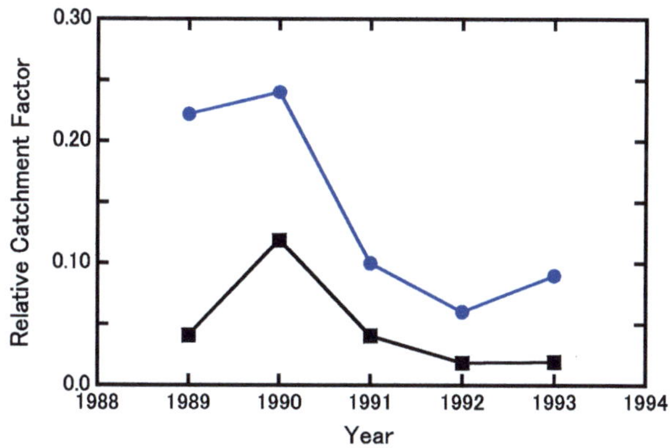

Fig. 6.2 Temporal changes in relative catchment factor ($RD_f(t)$) (*blue*; KR and *black*; KN)

progress of landform transformation, and it should be included in surface erosion models for appropriate evaluation, if some anthropogenic activities in the catchments affect erosion. Finally, the present observations show that the sedimentation rate is proportional to the rainfall intensity, and the catchment factor varies with the progress of landform transformation if the change in the pond factor is not so significant.

6.2 Experimental Models Based on Some Elementary Processes in the Lake-Catchment Systems

6.2.1 Outline of the Observed Systems

Mathematical expressions for various earth surface phenomena are the first step for a quantitative prediction. The reliability of the mathematical expressions depends on the degree of understanding of the processes underlying the targeted phenomena. Two pond-catchment systems in Japan and a system in Korea are selected to discuss an experimental model based on some elementary processes in this section (Kashiwaya et al. 2012).

The pond-catchment system in Japan is the Kawauso-ike system, which was often discussed in the previous sections. In addition to the earthquake and sedimentation studies, landslides and heavy rainfalls have also been investigated in this system (Kashiwaya et al. 1988, 1995, 2004). Observation of this system started just after the 1995 Kobe earthquake and continued until the end of 2007. Meteorological data obtained by the Kobe Marine Observatory were used for the hydrometeorological analyses of the system.

The Takidani-ike system is located in Kanazawa, Central Japan, facing the Japan Sea, where there is heavy snowfall during the winter (Figs. 6.3 and 6.4). Its catchment area covers 6.5 ha and has a relative relief of 100 m. Most of the catchment is covered with Japanese cedar. The average water area and the deepest

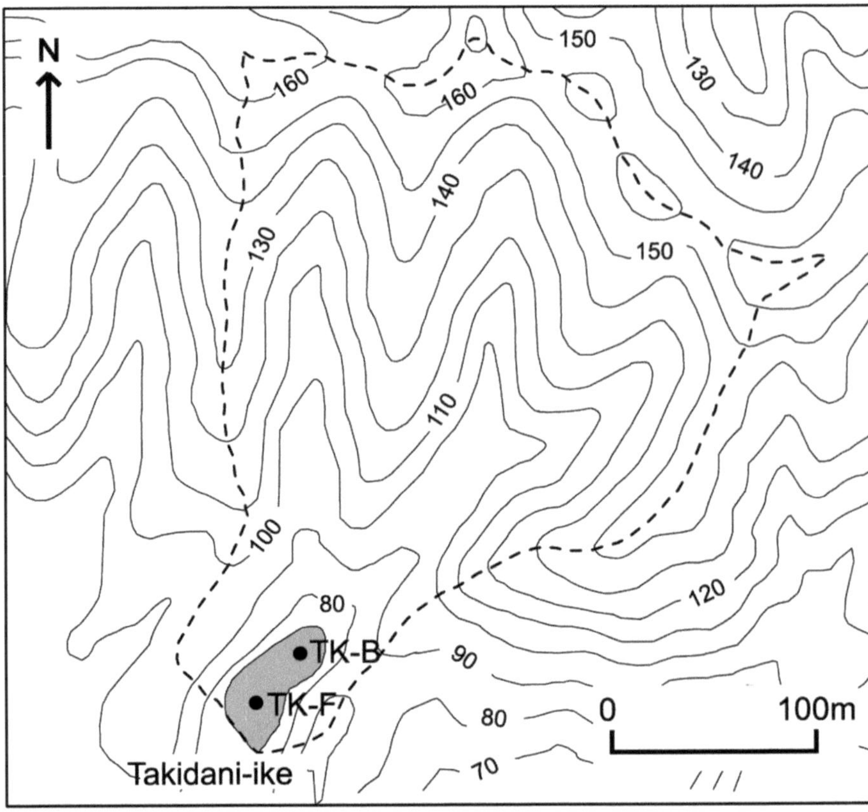

Fig. 6.3 Outline of the Takidani-ike system. Figures mean contour. TK-F and TK-B are sediment traps (modified after Ochiai et al. 2013)

point are approximately 0.24 ha and 5.0 m, respectively. The water level significantly changes during the summer irrigation season. The lithology around the pond and its catchment area is Miocene tuff. The reservoir has been used for irrigation of paddy fields, but it has not been dredged in the past several decades. There are two inflow channels and two artificial outflow channels. The upper part of the reservoir is often above water during water level low-stand periods. Two sediment traps were set at the bottom of the pond in May 2001 (TK-F and TK-B, Fig. 6.3), and the sediments gathered in the traps were sampled monthly. The water level was measured in the trap-sampling time before water level gauges in 2004. Meteorological data obtained by the Kanazawa Regional Observatory were used. Sediment transport processes in this system were also examined for checking fallout radionuclides due to the Fukushima Daiichi Nuclear Power Plant accident (Ochiai et al. 2015).

The Jinheung-je system is located in Jeongeup, Korea (Fig. 6.5), not so far from Pyeokgolje which was discussed before. The pond was also used for irrigation for more than two decades, but no dredging of the pond was reported (Nahm et al. 2010). Its catchment area covers 17.3 ha and has a relative relief of 100 m. The average

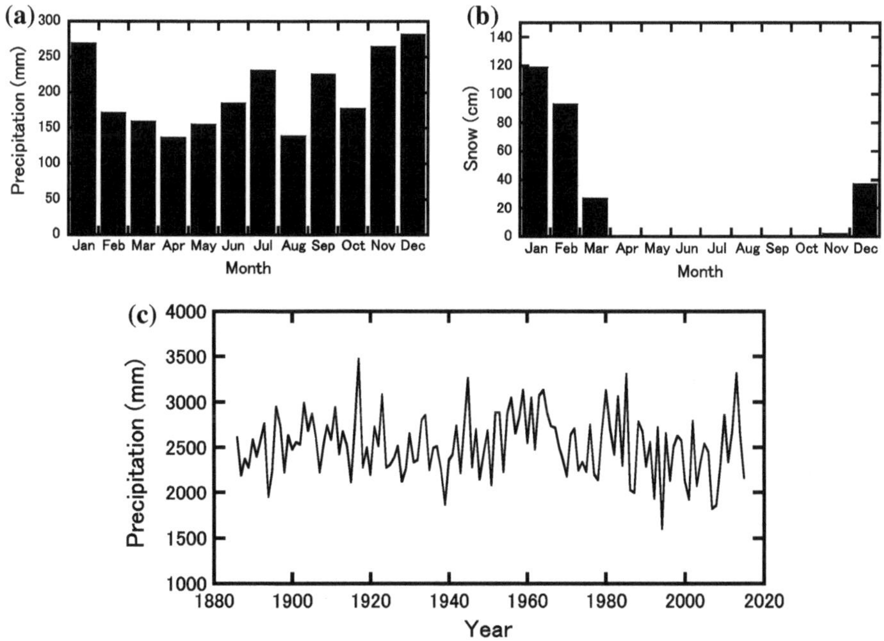

Fig. 6.4 Precipitation in Kanazawa Regional Observatory. **a** Monthly precipitation and **b** snow depth (averaged during the past 30 years; 1980–2010), and **c** annual precipitation

water area is approximately 0.89 ha, and its deepest point is approximately 2.2 m deep. The lithology around the pond and its catchment area is granite. The catchment area mainly comprises cayenne field with sparse *Pinus* woods. Two sediment traps were set in July 2003, and they were sampled monthly. Meteorological data obtained by the Kimje Meteorological Observatory were used for discussion.

The sizes of the three systems are roughly similar, and they are under similar climatic conditions (East Asian monsoonal zone), which is convenient for establishing a common model.

6.2.2 Experimental Models Based on Elementary Processes

An ideal simple process is introduced for material transportation in a lake-catchment system (Fig. 6.6). In general, temporal changes in an environmental factor (E) at a certain point (x, y) on the bottom in a lake-catchment system during a Δt interval at time t are expressed by (Kashiwaya et al. 2012):

$$\{E(t+\Delta t) - E(t)\}\Delta x\Delta y = \{J(x+\Delta x) - J(x)\}\Delta y\Delta t + \{J(y+\Delta y) - J(y)\}\Delta x\Delta t. \tag{6.5}$$

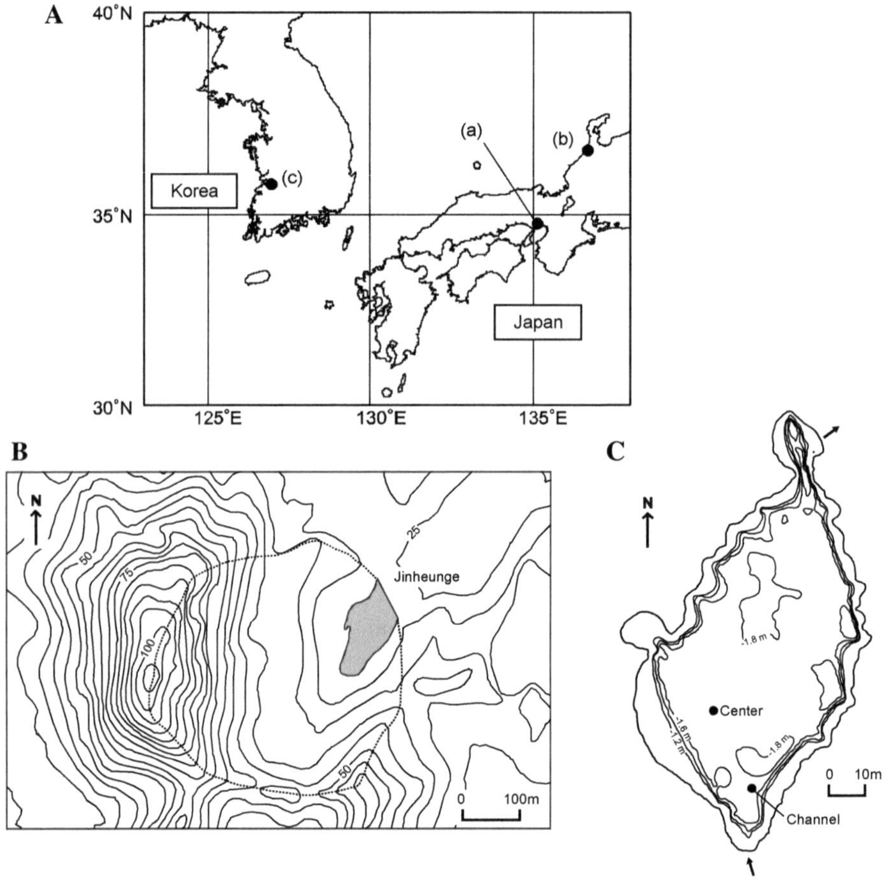

Fig. 6.5 **A** Location map for the three lake-catchment systems; (a) Kawauso-ike, (b) Takidani-ike, and (c) Jinheung-je, **B** Jinheung-je lake-catchment system, and **C** contour map in Jinheung-je and sediment traps (*closed circles*) (modified after Kashiwaya et al. 2012)

$$dE/dt = \partial J/\partial x + \partial J/\partial y, \tag{6.6}$$

where J is the flux of the factor, and the right side is given by:

$$\partial J/\partial x + \partial J/\partial y = Fe(C(t), D(t), L(t)), \tag{6.7}$$

where C refers to the climatic factors (precipitation, wind, temperature), D refers to the drainage (catchment) factors (landform, vegetation, transportable material, hydrological condition, etc.), and L refers to the lake factors (bed-form, transportable material, limnological condition, etc.).

Temporal changes in an environmental factor E_i in the *i-th* lake-catchment (region) can be expressed as:

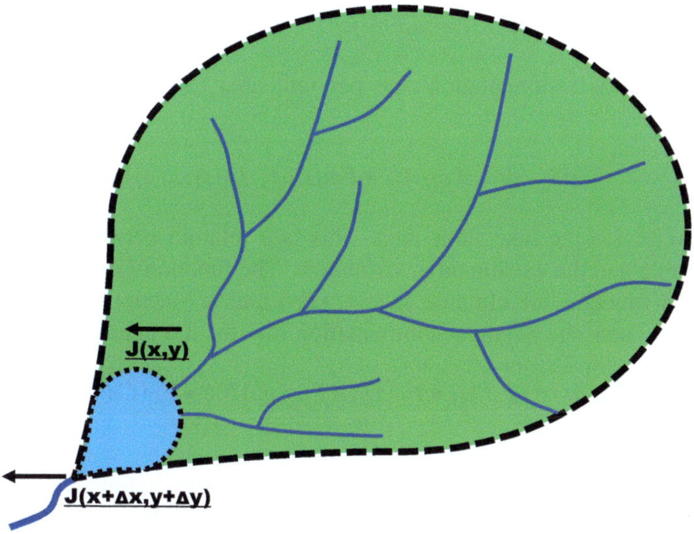

Fig. 6.6 An ideal model for a lake-catchment system

$$dE_i/dt = Fe_i(C_i(t), D_i(t), L_i(t)), \tag{6.8}$$

where dE, dt, C, D, and L are observables for a specific region (i) and time interval; the temporal changes of the environmental factors can be estimated for some regions in the instrumental observation period.

As a model case, let us discuss mass transport in a lake-catchment system. The temporal change in the volume of material moved in the catchment is:

$$dV_i/dt = F(C_i(t), D_i(t)), \tag{6.9}$$

where V_i, is the volume of material moved in the i-th region (catchment), C_i is the climatic factor, and D_i is the catchment factor in the i-th region. Here, $C_i(t) = f_{Ci}(T_i, R_i, W_i)$, where T_i is the temperature, R_i the precipitation (rainfall), and W_i the wind.

In the case of lake sediments, it is assumed that the sedimentation rate (dS_i/dt) is proportional to the material moved in the catchment.

$$dS_i/dt = \lambda_i \rho_i dV_i/dt, \tag{6.10}$$

where λ_i is a proportional factor related to the catchment factor D_i and the lake factor L_i [$\lambda_i = \lambda_i(D_i(t), L_i(t))$], and ρ_i is the sediment density. Therefore, Eq. (6.10) is expressed with a new function Fm of $C_i(t)$, $D_i(t)$, and $L_i(t)$,

$$\mathrm{d}S_i/\mathrm{d}t = \rho_i Fm(C_i(t), D_i(t), L_i(t)), \tag{6.11}$$

Then, the mass sedimentation rate per unit area ($S_{ai} = S_i/A_i$) is expressed as follows:

$$\mathrm{d}S_{ai}/\mathrm{d}t = (\rho_i/A_i) \bullet Fm(C_i(t), D_i(t), L_i(t)), \tag{6.12}$$

where A_i refers to the lake floor area. This is a relation between the mass sedimentation rate and the catchment environment. The function Fm is replaced by two functions: a function for climatic factors (F_C) and a function for other physical factors (F_P). Then, the average sedimentation rate for the interval (Δt) is

$$\Delta S_{ai}/\Delta t = (\rho_i/A_i) \bullet F_P(D_i(\Delta T), L_i(\Delta T)) \bullet F_C(C_i(\Delta T)), \tag{6.13}$$

here, $t + \Delta t \geq \Delta T \geq t$.

This is a simple form of the equation giving the sedimentation rate in a lake-catchment system. Then, it is necessary to provide functions available for the observational data in time and space, in order to check the model of the system. First, it is assumed that the catchment factor (D_i) and the lake factor (L_i) in the i-th region are constant for some intervals; $F_P(D_i(\Delta T), L_i(\Delta T)) = F_{Pi}$, and rainfall is a dominant climatic factor for erosion/transportation; $F_C(C_i(\Delta T)) = F_{Ri}(R_i(\Delta T))$, then

$$\Delta S_{ai}/\Delta t = (\rho_i/A_i) \bullet F_{Pi} \bullet F_{Ri}(R_i(\Delta T)). \tag{6.14}$$

Setting $SR_i = \Delta S_{ai}/\Delta t$, and $P_i = (\rho_i/A_i) \bullet F_{Pi}$, we obtain

$$SR_i = P_i \bullet F_{Ri}(R_{Si}), \tag{6.15}$$

where P_i are the physical environment factors characteristic for the i-th system, and $R_{Si} = R_i(\Delta T)$ is the specific rainfall related to mass transfer during the Δt interval for the i-th region. This equation may correspond to an experimental equation (model) for a simple lake-catchment process. P_i corresponds to $L_f(t) \bullet D_f(t)$ and $F_{Ri}(R_{Si})$ to $C_f(t)$ in the previous section. This means that the experimental equation introduced above can be also deduced from this process-oriented model.

6.2.3 Model Evaluation

The sedimentation rate (SR_i) is expressed as a function of the physical environment factors (P_i) and the rainfall intensity (R_{Si}), as shown in Eq. (6.15). The factor (P_i) characteristic for a lake-catchment system is assumed approximately constant for a certain interval. Then, the appropriate sedimentation rate and the rainfall intensity are required to validate the model for this interval. The relative sedimentation rate

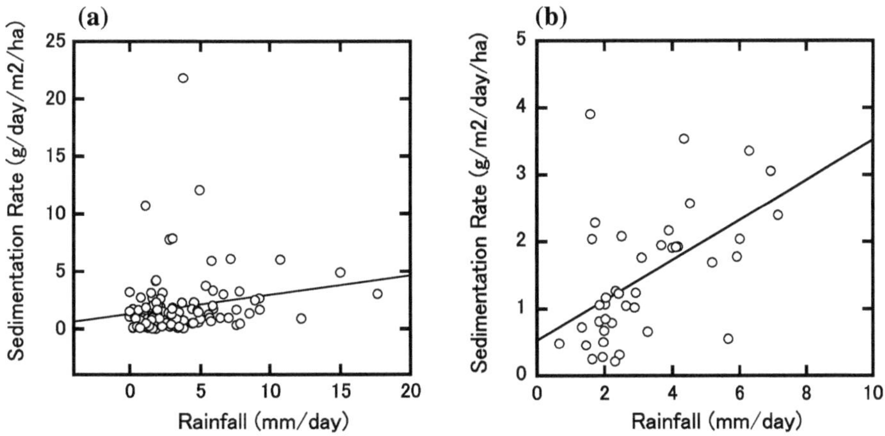

Fig. 6.7 **a** Relationship between monthly rainfall (mm/day) and monthly sedimentation rate (g/m^2/day/ha), and **b** relationship between seasonal rainfall (mm/day) and seasonal sedimentation rate (g/m^2/day/ha) in Kawauso-ike (modified after Kashiwaya et al. 2012)

(the average sedimentation rate divided by the catchment area) is used for the three systems of different lake-catchment scales.

The Kawauso-ike system was used for monitoring environmental changes since the 1995 Kobe earthquake. As indicated in the previous section, the sedimentation rate abruptly increased just after the earthquake and has gradually decreased since then, and the seasonal rainfall was roughly proportional to the seasonal sedimentation rate (Kashiwaya et al. 2004). Here, two sets of observational data are shown (the monthly sedimentation rate and rainfall, and the seasonal sedimentation rate and rainfall), which suggest that the rainfall is correlated better with the seasonal rate than with the monthly rate (Fig. 6.7). As indicated before, after the earthquake, in the earlier half of the observation period, the seasonal sedimentation rate gradually decreased, roughly following the rainfall. However, it seems that there is no gradually decreasing trend in the latter half, although the rate is proportional to the rainfall (Fig. 6.8), suggesting a change in the pond-catchment processes introduced later. The gradual decrease in the earlier half is explained that surface fine materials produced during the earthquake time were gradually transported with rainfall, and then a different state might be established (Kashiwaya et al. 2015). The seasonal sedimentation rate in the latter half is more clearly proportional to the seasonal rainfall than that in the earlier half, indicating that enough materials were eroded by rainfall (Fig. 6.9a). The surface materials of this area comprise deeply weathered granite, which caused frequent landslides due to periods of heavy rainfall in the past (Kashiwaya et al. 1988, 1995). Most of the surface fine materials added by the earthquake-landslides were probably transported from the catchment in the earlier half of the observation period, and the skeleton-destroyed weathered granites were mainly eroded in the latter half (Kashiwaya et al. 2015).

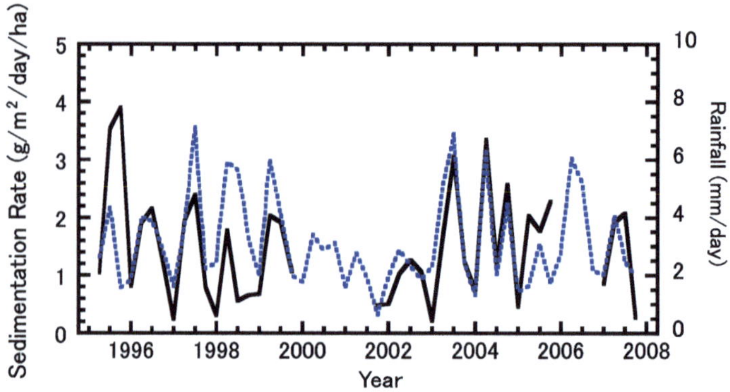

Fig. 6.8 Temporal changes in seasonal sedimentation rate (g/m^2/day/ha) (*solid line*) and seasonal rainfall (mm/day) (*dotted line*) in Kawauso-ike (modified after Kashiwaya et al. 2012)

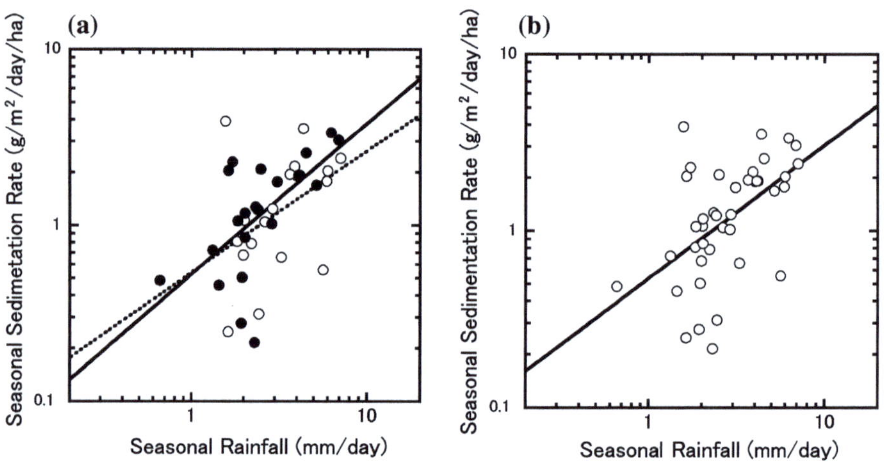

Fig. 6.9 **a** Relationship between seasonal rainfall (mm/day) and seasonal sedimentation rate (g/m^2/day/ha) in the two intervals. *Open circles* and *dotted regression curve*; 1995–2000, $Y = 0.54X^{0.70}$ ($R = 0.31$), *closed circles* and *solid regression curve*; 2001–2008, $Y = 0.53X^{0.86}$ ($R = 0.76$) and **b** relationship between seasonal rainfall (mm/day) and seasonal sedimentation rate (g/m^2/day/ha) in Kawauso-ike for the whole interval; $Y = 0.54X^{0.75}$ ($R = 0.58$). X; seasonal rainfall, Y; seasonal sedimentation rate

The relation between the seasonal sedimentation rate and rainfall in the whole interval is shown in Fig. 6.9b. The Eq. (6.15) is applicable to this pond-catchment system because the sedimentation rate, especially the seasonal sedimentation rate, seems to be a function of seasonal rainfall. In this case, Eq. (6.15) is generally expressed as

$$SR_i = P'_i \bullet (R_{Si})^b, \tag{6.16}$$

where P'_i is a physical environment factor related to transportability (system movable factor), and b is a hydrological factor related to the runoff properties of the system structure (runoff factor). A regression curve corresponding to this equation is shown in Fig. 6.9b (solid line). Moderate correlation here suggests that the system movable factor is not always constant with time.

The Takidani-ike system in Kanazawa, central Japan, has two specific characteristics affecting the hydrological conditions: snow cover during the winter and artificial water level change due to irrigation in the summer. Temporal changes (monthly and seasonal) in the sedimentation rate and precipitation (rainfall) are shown in Fig. 6.10, indicating that the sedimentation rate does not seem to be related to precipitation (rainfall), although the seasonal sedimentation rate seems to correspond to the precipitation in some intervals except during the winter period. This is mainly attributed to the winter snow cover and the artificial water level fall due to irrigation. The relation between the seasonal sedimentation rate and precipitation

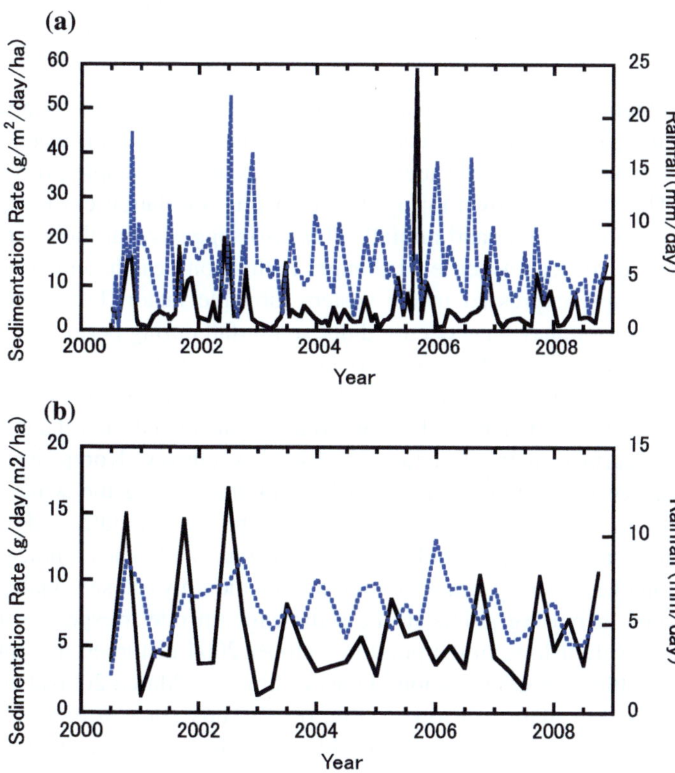

Fig. 6.10 Temporal changes in **a** monthly sedimentation rate (g/m^2/day/ha) (*solid line*) and monthly rainfall (mm/day) (*dotted line*) and **b** seasonal sedimentation rate (g/m^2/day/ha) (*solid line*) and seasonal rainfall (mm/day) (*dotted line*) in Takidani-ike (modified after Kashiwaya et al. 2012)

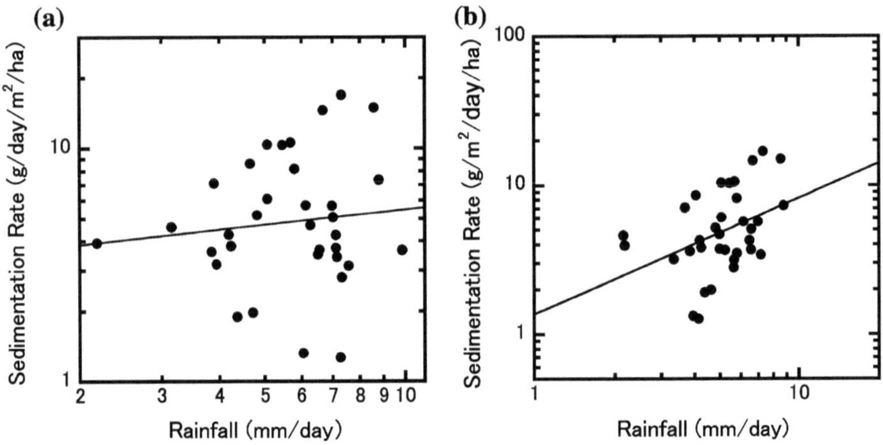

Fig. 6.11 **a** Relationship between seasonal rainfall (mm/day) and seasonal sedimentation rate (g/m^2/day/ha) in the whole interval (*closed circles* and *solid regression curve*) (regression curve; $Y = 3.31X^{0.22}$ ($R = 0.19$)), and **b** relationship between seasonal rainfall (mm/day) and seasonal sedimentation rate (g/m^2/day/ha) in Takidani-ike for the interval without winter season (regression curve; $Y = 1.36X^{0.78}$ ($R = 0.47$)). X; seasonal rainfall, Y; seasonal sedimentation rate

(rainfall) including all observational data, as shown in Fig. 6.11a, is not so good, probably reflecting the strong effect of snow cover and artificial water level fall. Then, intervals with snow cover are excluded to examine a simple process without this effect. The result is shown in Fig. 6.11b. A fair correlation exists between the seasonal sedimentation rate and rainfall. This also suggests that there remain some effects related to artificial water level changes in this pond-catchment system, which may not be simply expressed in the system movable factor. For example, pond bottom-floor erosion may have occurred due to rainfall during periods of significant artificial water level lows. Lake (pond) floor exposure prompts the erosion and transportation of surface materials in the floor more easily. This may give a hint to recognize water level changes and erosion during glacial and interglacial periods.

The water surface in the Jinheung-je system, southwest Korea, is completely covered with ice in the winter. There is little rainfall during the winter and little sedimentation from outside the system. Temporal changes (monthly and seasonal) in the sedimentation rate and rainfall are shown in Fig. 6.12. These indicate that the sedimentation rate is closely related to rainfall both monthly and seasonally, although there are some increasing points in the sedimentation rate irrespective of rainfall values. Large-scale fishing in the pond in October 2004 disturbed the bottom sediments, and a cattle-shed construction in the catchment in March 2006 also prompted an increase in sedimentation (arrows in Fig. 6.12). Thus, some artificial events were also recorded clearly in the sediments (Kashiwaya et al. 2012). The relations between the monthly and seasonal sedimentation rates and rainfall (except for the periods of artificially induced rapid sedimentation rates) are shown in Fig. 6.13, indicating that the seasonal sedimentation rate is correlated better with the seasonal rainfall

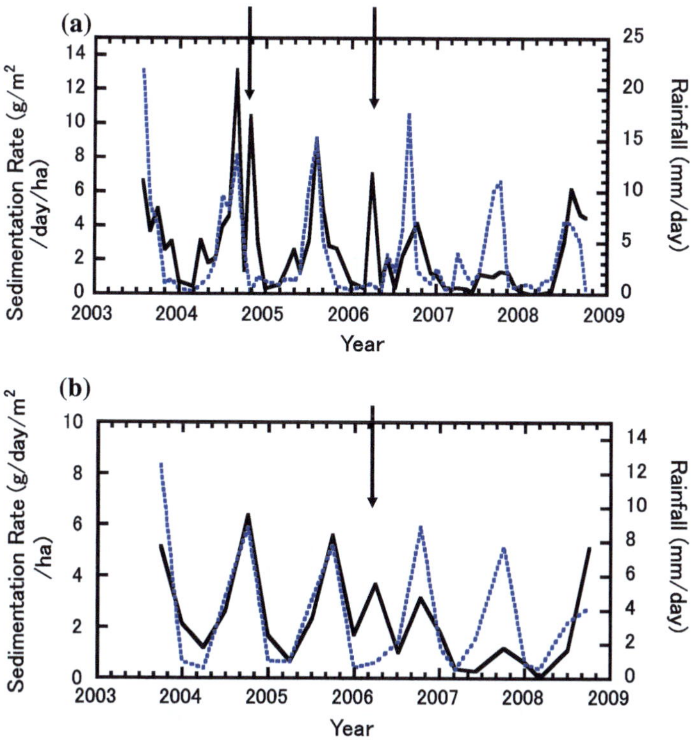

Fig. 6.12 Temporal changes in **a** monthly sedimentation rate (g/m^2/day/ha) (*solid line*) and monthly rainfall (mm/day) (*dotted line*) and **b** seasonal sedimentation rate (g/m^2/day/ha) (*solid line*) and seasonal rainfall (mm/day) (*dotted line*) in Jinheung-je (modified after Kashiwaya et al. 2012). *Arrows* mean artificial events

(Fig. 6.13b) and is more correlative than the monthly sedimentation is (Fig. 6.13a). This is also related to the reservoir effect in the catchment.

As shown above, there is a reasonable (good/fair) relation between the seasonal sedimentation rate and the seasonal rainfall in the three systems. Then, let us discuss the relationship using Eq. (6.15). The above calculated results show that Eq. (6.16) using power functions for seasonal factors seems to be more appropriate for present lake-catchment systems. The relation between the seasonal rainfall and the sedimentation rate for the three systems is shown in Fig. 6.14 (SR_i; seasonal sedimentation rate and R_{Si}; seasonal rainfall), supporting that the above model (Eqs. (6.15) and (6.16)) is acceptable as a first approximation. Regression lines, expressed with power functions for each site, are also shown in the figure, indicating that the power of the functions (b), defined as the runoff factor, is nearly the same (0.75–0.78). This may be closely related to lake-catchment conditions although not all regressions show good correlation coefficients. This suggests that the hydrogeomorphological structure of the pond-catchment system is similar in the three systems. The physical environment factor (P_i') slightly varies between the

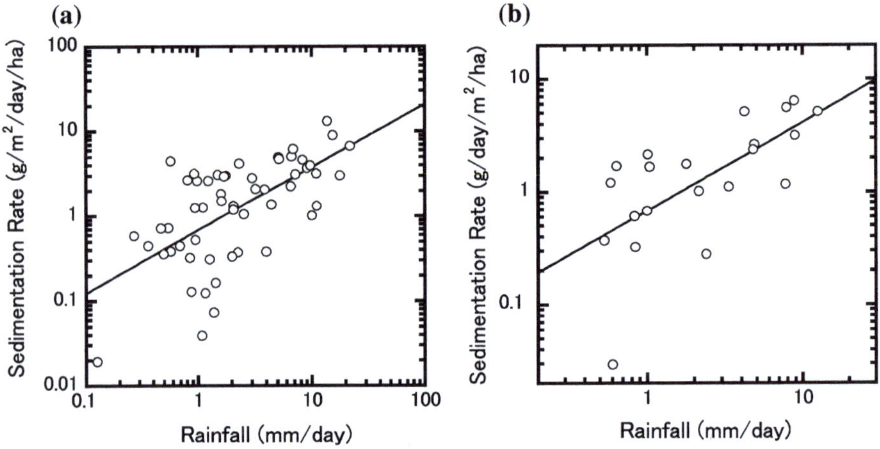

Fig. 6.13 **a** Relationship between monthly rainfall (mm/day) and monthly sedimentation rate (g/m^2/day/ha) in the interval without artificial events (*open circles* and *solid regression curve*; $Y = 0.68X^{0.74}$ ($R = 0.66$)) and **b** relationship between seasonal rainfall (mm/day) and seasonal sedimentation rate (g/m^2/day/ha) in Jinheung-je without artificial events (*open circles* and *solid regression curve*; $Y = 0.54X^{0.78}$ ($R = 0.76$)). X; rainfall, Y; sedimentation rate

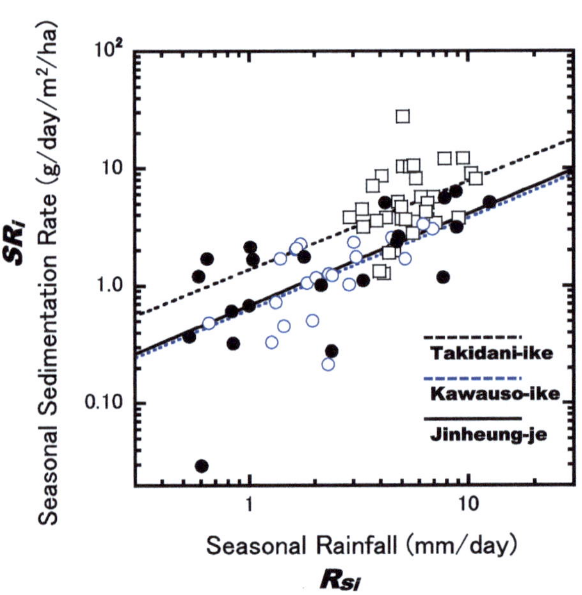

Fig. 6.14 Relationship between seasonal rainfall (mm/day) and seasonal sedimentation rate (g/m^2/day/ha) for the three systems. *Open squares* and *black dotted line* ($P_i' = 1.36$, $b = 0.78$); Takidani-ike, *open circles* and *blue dotted line* ($P_i' = 0.54$, $b = 0.75$); Kawauso-ike, and *solid circles* and *black line* ($P_i' = 0.68$, $b = 0.78$); Jinheung-je

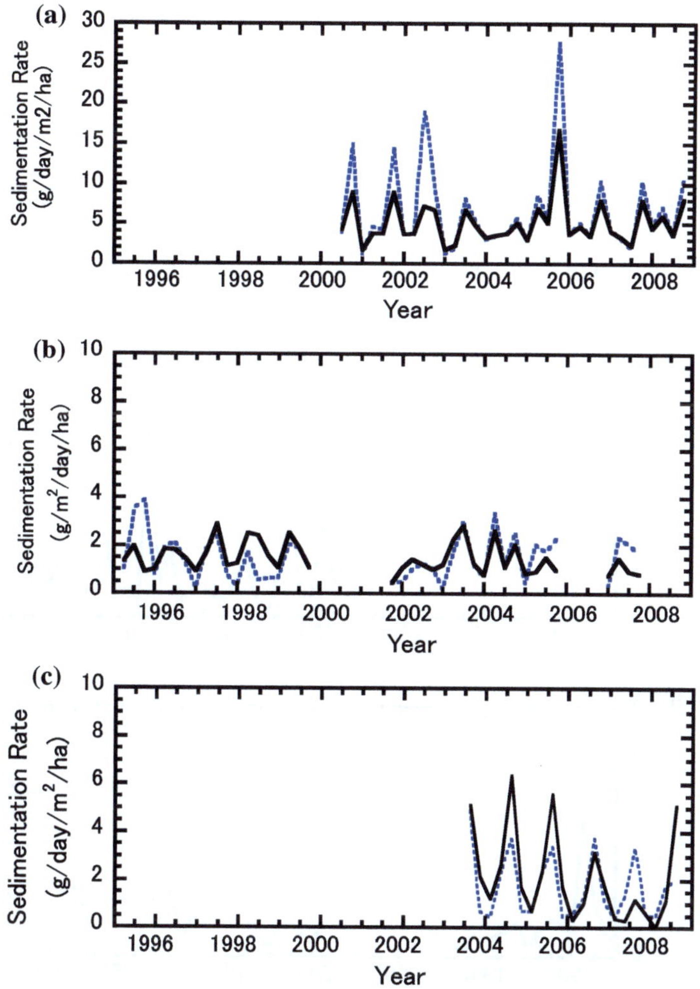

Fig. 6.15 Temporal changes in sedimentation rate for the three systems **a** Takidani-ike, **b** Kawauso-ike, and **c** Jinheung-je. *Solid line*; calculated, *dotted line*; observed

three systems. In the Takidani-ike system, it is the largest among the three, indicating that the surface materials of the Takidani-ike system are more easily moved than those of the Kawauso-ike and the Jinheung-je systems are. The Kawa-ike (KR and KN) systems in Kobe discussed above have a similar hydrogeomorphological structure to those of these three systems (the values for the factor b are 0.73–0.75, although they are given for the monthly interval; Fig. 6.1).

Using the two factors, we can estimate the temporal changes in the sedimentation rate and compare them with observed ones, shown in Fig. 6.15, indicating that the basic equations for the model are expressed in Eqs. (6.15) or (6.16). If the relation shown in Eq. (6.16) expresses the basic process, we may evaluate a time series of the physical environment factor (system movable factor) P_i' because SRi and $(R_{Si})^b$ are given from field observations;

$$P_i' = SRi/(R_{Si})^b \tag{6.17}$$

The calculated results for the three systems are shown in Fig. 6.16, which shows that the values are large during the summer and autumn and small in the winter seasons in all systems (changes in the factor are clear in Takidani-ike and Jinheung-je, although they are not so clear in Kawauso-ike), reflecting a seasonal

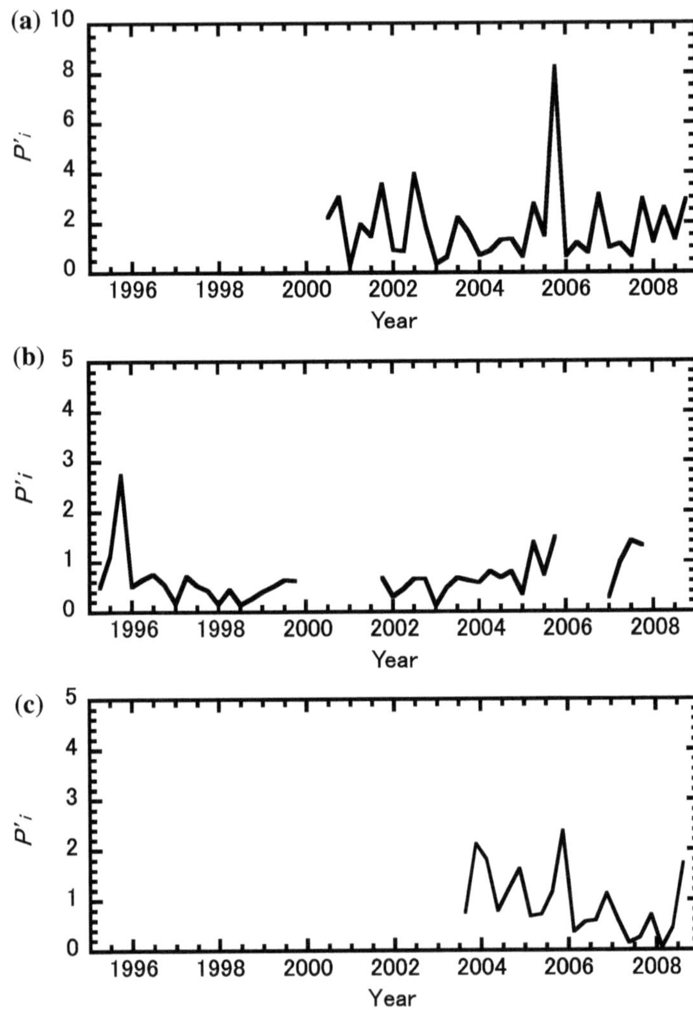

Fig. 6.16 Temporal changes in the physical environment factor (system movable factor) P_i' for the three systems **a** Takidani-ike, **b** Kawauso-ike, and **c** Jinheung-je

influence on the factor. During the winter, the water surface is covered with ice/snow in Takidani-ike and Jinheung-je, but the ice/snow coverage is less in Kawauso-ike. Takidani-ike has the largest values of the factor with a large amplitude, and Kawauso-ike has the smallest factor with a small amplitude. This is partially explained by the difference in the catchment inclination; the average inclination in the catchment is 19.6° for the Kawauso-ike system and 31.5° for the Takidani-ike system, and the weathered (erodible) layer of the catchment in Takidani-ike is larger than in Kawauso-ike. The weathered layer here is defined as the penetrated depth with a 5-kg-weight knocking (the layer where the 10-cm penetration depth is given under 10 knocks) (Kashiwaya et al. 2012). Large values for the Takidani-ike system in summer seasons are also related to water level change (for irrigation), as discussed above. The values for the Kawauso-ike system are comparatively small except those corresponding to the earlier interval, when the 1995 Kobe earthquake may have influenced the factor. The relative catchment factor discussed above for Kawa-ike corresponds to this system movable factor, although the values are yearly averaged ones (Fig. 6.2).

These results lead to that the basic model, where the sedimentation rate is a function of the rainfall intensity and the system conditions (system movable factor), is acceptable as the first step toward understanding the underlying processes in the lake-catchment system based on field data. The basic concept discussed here can be extended if the system conditions are clarified and expressed with mathematical formulations, as is going to be discussed in the next section. Additionally, it is important to consider other environmental conditions with appropriate modifications, as for example, where surface water discharge due to melting snow and/or ice as well as rainfall is a dominant external force. This also contributes to connecting environmental processes with temporal (historical) changes using long observational data.

References

Jones DS, Kowalski DG, Shaw RB (1996) Calculating revised Universal Soil Loss Equation (RUSLE) estimates on department of defense lands: a review of RUSLE factors and U.S. Army land condition—trend analysis (LCTA) data gaps. Department of Forest Science, Colorado State University, Fort Collins, Colorado, 9p

Kashiwaya K, Taishi H, Kawatani T, Okimura T (1988) Grain size variation of pond sediments and landslide environment in the Rokko Mountains. Trans Jpn Geomorphol Union 9:193–200 (in Japanese with English abstract)

Kashiwaya K, Okimura T, Kawatan T, Aoki T, Isozumi Y (1995) Surface erosional environment and pond sediment information. In: Slaymaker O (ed) Steepland Geomorphology. Wiley, Chichester, 283p, pp 218–231

Kashiwaya K, Okimura T, Harada T (1997) Land transformation and pond sediment information. Earth Surf Proc Land 22:913–922

Kashiwaya K, Tsuya Y, Okimura T (2004) Earthquake-related geomorphic environment and pond sediment information. Earth Surf Proc Land 29:785–793

Kashiwaya K, Ochiai S, Okimura T, Nahm WK, Yang DY, Kim JY (2012) Erosion and sedimentation in lake-catchment systems in Japan and Korea on the basis of an elementary process model. Trans Jpn Geomorphol Union 33:219–236

Kashiwaya K, Okimura T, Itono T, Ishikawa K, Kusumoto T (2015) Present earth-surface processes and historical environmental changes inferred from lake-catchment systems. In: Kashiwaya K, Shen J, Kim JY (eds) Earth surface processes and environmental changes in East Asia—records from lake-catchment systems. Springer, Tokyo, 321p, pp 1–24

Nahm WH, Lee GH, Yang DY, Kim JY, Kashiwaya K, Yamamoto M, Sakaguchi A (2010) A 60-year record of rainfall from the sediments of Jinheung Pond, Jeongeup, Korea. J Paleolimnol 43:489–498

Ochiai S, Nagao S, Yamamoto M, Itono T, Kashiwaya K, Fukui K, Iida H (2013) Deposition records in lake sediments in western Japan of radioactive Cs from the Fukushima Dai-ichi nuclear power plant accident. Appl Radiat Isot 81:366–370

Ochiai S, Yamamoto M, Nagao S, Itono T, Kashiwaya K (2015) Sediment transport processes in a reservoir–catchment system inferred from sediment trap observations and fallout radionuclides. J Radioanal Nucl Chem 303:1497–1501

Chapter 7
Observation of a Small Lake-Catchment System (Kawauso-ike System) After the Kobe Earthquake and Mathematical Models

Various forces acting on lake-catchment systems have been recorded in lacustrine sediments as mentioned above. Thus, it is necessary to quantitatively express temporal changes in physical phenomena to understand the lake-catchment processes. This will allow the provision and prediction as well as the appropriate interpretation of past processes and changes. Observation of the Kawauso-ike system after the 1995 earthquake started in May 1995. Here, a mathematical model of the temporal changes in the system is introduced with more than 10-year observations, although a basic idea of the model on the observation results before 2000 was introduced by Kashiwaya et al. (2004) and a modified one was proposed by Kashiwaya et al. (2015). This section is mainly based on this modified idea.

Two sediment traps were placed on the pond bed in May 1995 (two circles in the pond, Fig. 4.24). Sediments were picked up from the traps once or twice a month, and the sediments' physical properties were measured. Meteorological data were obtained for discussion from the Kobe Marine Observatory.

Figure 4.26a shows the temporal changes in the median grain size of sediments obtained with traps for the earlier five years (the first stage). Overall, only a small difference is observed in the samples, despite small fluctuations in recent years. This may reflect the fact that there were few instances of extremely heavy rainfall and rapid mass movements during this interval (Fig. 4.26b).

The annual sedimentation rate inferred from the two sediment traps is shown in Fig. 4.25a. The rate was high just after the earthquake and then gradually decreased. Seasonal changes in the sedimentation rate are generally controlled by seasonal rainfall, but no large fluctuation is observed in the annual (seasonal) rainfall, despite the temporal decrease in the sedimentation rate (Fig. 4.25b). Therefore, the influence of the earthquake on the sedimentation rate in the second stage (2002–2007) does not seem significant, although the rate was higher than in the pre-earthquake period (Fig. 4.25a). The discussion on transportability (system movable factor) in the previous section supports this conclusion. Torii et al. (2007) reported that the increase in material erodibility after the earthquake was related to the destruction of the skeletal structure of weathered materials (granites) with

K. Kashiwaya, *Geomorphology of Lake-Catchment Systems*,
Environmental Earth Sciences, DOI 10.1007/978-981-10-5110-4_7

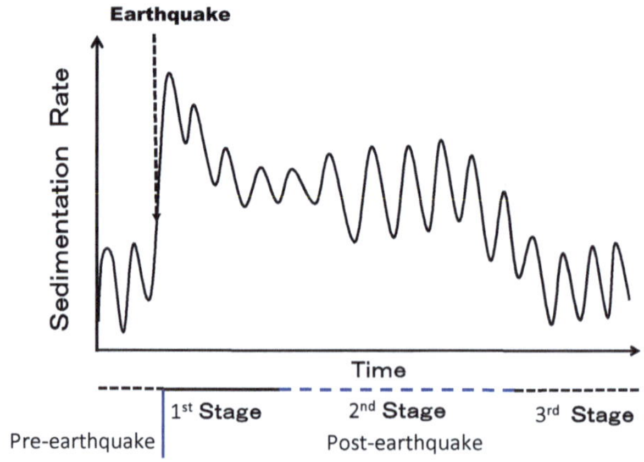

Fig. 7.1 A basic idea on temporal change in the sedimentation rate pre- and post-earthquake and sedimentary stages (modified after Kashiwaya et al. 2015)

earth-quaking in addition to fine surface materials due to landslides in the earthquake. Thus, three stages may be assumed for establishing a model after the earthquake: (1) rapid increase and gradual decrease in sedimentation rate, (2) stationary stage in sedimentation rate and higher erodibility than before the earthquake, and (3) new weathering-limited stage, similar to that before the earthquake. The temporal changes in the sedimentary rate and the sedimentary stages around the earthquake may be considered as shown in Fig. 7.1. An ideal longitudinal state in the catchment after the earthquake is assumed, as shown in Fig. 7.2. In the case of Kawauso-ike, two stages (the first stage and the second stage) are available for discussion with field observation and data.

Surface fine materials produced in the earthquake should be eroded and transported downstream mainly by rainfall in the first stage. The basic sedimentation rate for the *i*th lake-catchment system is given by Eqs. (6.9) and (6.10) in the previous

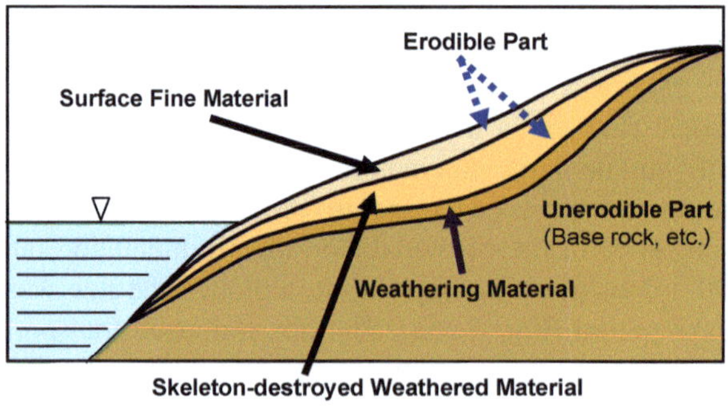

Fig. 7.2 A conceptual model of erosion–sedimentation process in a lake-catchment system after an earthquake (modified after Kashiwaya et al. 2015)

section. It is assumed that the sedimentation rate in the lake ($SR(t)$ = d$S(t)$/dt) is proportional to the erosion rate in its catchment (d$M(t)$/dt):

$$dS(t)/dt = \alpha(t)dM(t)/dt \tag{7.1}$$

where $M(t)$ is the volume of material to be moved at time t and $\alpha(t)$ is a proportionality factor (sediment delivery factor related to system conditions) in the catchment. The erosion rate in the catchment at t is proportional to the volume of material to be eroded ($V(t)$):

$$dM(t)/dt = \lambda(t)V(t), \tag{7.2}$$

where $\lambda(t)$ is the relative erosion factor related to external forces and system/material conditions (e.g., rainfall intensity, geomorphic settings, and physical properties).

It is assumed that only surface fine materials produced in the earthquake are transported in the first stage. The volume of fine materials transported at time t ($V(t)$) is expressed as follows:

$$V(t) = V_T - \{V_{DW} + V_W(t) + V_U(t)\} - M(t), \tag{7.3}$$

where V_T is the total volume of material in a catchment area at the initial stage, V_{DW} is the skeleton-destroyed weathered material, $V_W(t)$ is the weathering material, and $V_U(t)$ is the unerodible part at time t.

Then, we obtain:

$$-\frac{dV(t)}{dt} = \lambda(t)V(t) + \frac{d\{V_W(t) + V_U(t)\}}{dt}. \tag{7.4}$$

Replacing $d\{V_W(t) + V_U(t)\}/dt = f_{WU}(t)$, Eq. (7.4) can be solved and we obtain the sedimentation rate as follows:

$$SR(t) = \alpha(t)\lambda(t)V(t) = \alpha(t)\lambda(t)\left[e^{-\int^t \lambda(\tau)d\tau}\left\{-\int^t f_{WU}(t)e^{-\int^t \lambda(\tau)d\tau}dt + V_0\right\}\right], \tag{7.5}$$

with $V(0) = V_0$ (initial surface fine materials). As $\lambda(t)$ is a factor related to external forces and system (material) conditions (e.g., rainfall intensity, geomorphic settings, and physical properties), it may be expressed as a product of the rainfall factor function $\beta(t)$ and the non-rainfall factor function $\gamma(t)$ for appropriate use of the field data:

$$\lambda(t) = \beta(t)\gamma(t).$$

As mentioned above, it is assumed that only surface fine materials are eroded in the first stage because no extremely heavy rainfall occurred since the 1995 Kobe earthquake. If changes in the weathered part and the unerodible part are limited in the short interval ($V_W(t)$, $V_U(t) \approx$ const), $f_{WU}(t)$ may be considered zero. Then, Eq. (7.5) becomes:

$$SR_1(t) = \alpha_1(t)\lambda_1(t)V_{01}e^{-\int^t \lambda_1(\tau)d\tau}, \tag{7.6}$$

where $\alpha_1(t)$ is a proportionality factor (sediment delivery factor for the first stage) related to the surface fine materials, $\lambda_1(t)$ is a relative erosion factor, and V_{01} is the initial volume of material in the first stage (surface fine materials). If changes in the sediment delivery factor and the non-rainfall factor are limited for a certain interval, the model can be checked with an appropriate function for $\lambda_1(t) = \gamma_1\beta_1(t)$ and appropriate values of α_1, γ_1, and V_{01} using field data. The rainfall factor $\beta_1(t)$ is assumed a temporal function of seasonal rainfall, which is given through Fourier approximation by:

$$\beta_1(t) = \sum_i p_{1i}\sin q_{1i}(t - t_{1i}) + r_{1i}. \tag{7.7}$$

The calculated result for the factor is shown in Fig. 7.3a (the solid curve is the original seasonal rainfall, and the dotted curve is the synthesized seasonal rainfall).

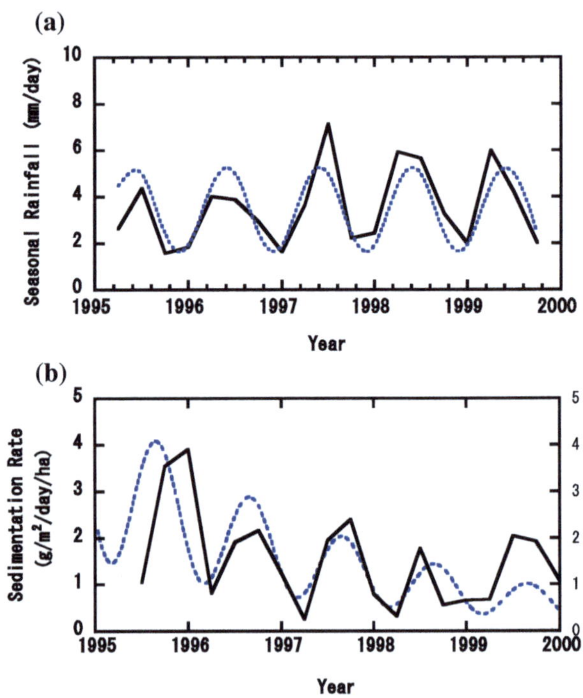

Fig. 7.3 **a** Changes in the seasonal rainfall (mm/day) for the first stage. *Solid line* is observed and *dotted line* is synthesized. **b** Changes in the seasonal sedimentation rate (g/cm^2/day/ha) in the lake-catchment system. *Solid line* is observed and *dotted line* is calculated (modified after Kashiwaya et al. 2015)

The theoretical sedimentation rate using Eq. (7.6) (dotted line) is shown in Fig. 7.3b with the observational curve (solid line). This indicates that the model introduced here is valid for the erosion-related sedimentation processes in the first stage after the earthquake.

In the second stage (2002–2008), the sedimentation rate appears to be controlled by the rainfall intensity (seasonal sedimentation rate and seasonal rainfall). The values are little larger than those before the earthquake. This suggests that the catchment is under more erodible conditions than before the earthquake. This is possibly related to the skeleton-destroyed weathered materials produced by the earthquake (Torii et al. 2007); the earthquake probably continues to influence the earth-surface environment, although its direct influence rapidly diminishes (most earthquake-produced fine materials in the catchment are smoothly transported downstream). In this stage, most materials to be eroded are the skeleton-destroyed weathered granite (Fig. 7.2; probably, the structure of the surface weathered layer is weakened by the earthquake). In the short interval of the second stage, the term ($f_{WU}(t)$) for further weathering and uplifting (increase in unerodible part) in Eq. (7.4) may be neglected. Then,

$$SR_2(t) = \alpha_2 \lambda_2(t) V_{02} \mathrm{e}^{-\int^{t} \lambda_2(\tau)\mathrm{d}\tau}, \tag{7.8}$$

where α_2 is a proportionality factor related to the skeleton-destroyed materials, $\lambda_2(t)$ ($= \gamma_2\beta_2(t)$) is a relative erosion factor in the second stage, and V_{02} is the initial volume of material in the second stage (skeleton-destroyed weathered materials). The rainfall factor $\beta_2(t)$ is also assumed a temporal function of seasonal rainfall:

$$\beta_2(t) = \gamma_2 \sum_i p_{2i} \sin q_{2i}(t - t_{2i}) + r_{2i}. \tag{7.9}$$

Using a Fourier approximation, this equation is expressed using the seasonal rainfall data (dotted line in Fig. 7.4a). The theoretical sedimentation rate calculated using Eq. (7.8) (dotted line) and the observations (solid line) are shown in Fig. 7.4b, indicating that the model is also valid for the second stage when skeleton-destroyed materials are mainly eroded. The results obtained here suggest that the presented model is acceptable for the erosion–sedimentation processes in the small lake-catchment system for the earthquake-influencing interval (first and second stages).

In general, earthquakes produce a large yield of fine material in addition to significant coarse material due to landslides in mountainous areas. Surface fine materials are easily moved even by a small rainfall. This explains the rapid increase in the rate of deposition of fine materials just after the earthquake. The 1995 Kobe Earthquake also produces skeleton-destroyed weathered materials, as indicated by the observational data; their sedimentation rate becomes comparatively stable after a period of rapid increase and gradual decrease, larger than the rate before the earthquake. The model introduced here is based on simple assumptions. Thoroughly investigating changes in lake-catchment conditions and processes related to earthquakes, hydrological conditions, and weathering is necessary to

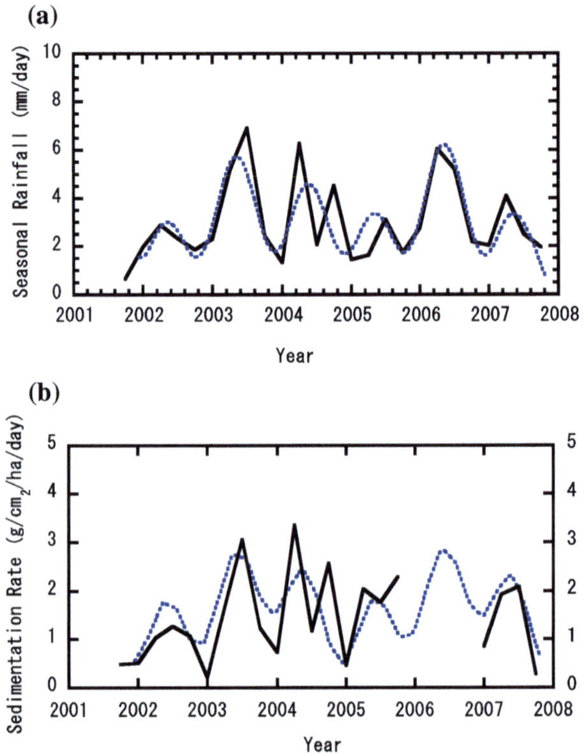

Fig. 7.4 **a** Changes in the seasonal rainfall (mm/day) for the second stage. *Solid line* is observed and *dotted line* is synthesized. **b** Changes in the seasonal sedimentation rate (g/cm^2/day/ha) in the lake-catchment system. *Solid line* is observed and *dotted line* is calculated (modified after Kashiwaya et al. 2015)

make the model more physically meaningful. Nevertheless, these equations will be available for the quantitative prediction of erosion–sedimentation in near future (10^1–10^2 years), if the physical properties α_1 and α_2 are properly obtained from field data and material conditions as well as appropriate values of V_{DW}, V_{01}, V_{02}, etc.

References

Kashiwaya K, Tsuya Y, Okimura T (2004) Earthquake-related geomorphic environment and pond sediment information. Earth Surf Proc Land 29:785–793

Kashiwaya K, Okimura T, Itono T, Ishikawa K, Kusumoto T (2015) Present earth-surface processes and historical environmental changes inferred from lake-catchment systems. In: Kashiwaya K, Shen J, Kim JY (eds) Earth surface processes and environmental changes in East Asia—records from lake-catchment systems. Springer, Tokyo 321p, pp 1–24

Torii N, Okimura T, Kato S (2007) Experimental study on slope failure mechanism induced by rainfall after earthquake. J JSCE (C) 63(1):140–149 (in Japanese with English abstract)

Chapter 8
Long-Term Changes and Phenomenological Models

8.1 Background for Establishing Models

As discussed in the previous chapters, the lake-catchment systems of Lake Biwa (Figs. 3.1, 3.2, and 3.3), Lake Baikal (Figs. 3.4, 3.5, 3.6, 3.7, 3.8, 3.9, and 3.10), and Lake Khuvsgul (Figs. 3.11, 3.12, 3.13, and 3.14) are of great use for reconstructing long-term environmental changes, although there are large differences in the environmental conditions between Lake Baikal and Lake Khuvsgul, and Lake Biwa. There were large glaciers during glacial periods in the catchments of two continental lakes, whereas there was no glacier in the catchment of Lake Biwa.

Changes in physical earth-surface environments of catchments may be recorded in sediments flowing into lakes. Mineral grain size is often used as a proxy for sediment discharge and/or precipitation. As shown in Figs. 3.2 and 3.3, changes in the bi-SiO_2 content (proxy for temperature) and the mineral grain size (proxy for precipitation and/or sediment discharge) indicate that regional hydrogeomorphological fluctuations in Lake Biwa were roughly controlled by global climatic changes, although some regional phenomena significantly influenced the hydrogeomorphological conditions, especially the grain size fluctuation.

Long records from Lake Baikal show that changes in the bi-SiO_2 content and the mineral content are closely related to global climatic changes as expressed by the $\delta^{18}O$ concentration in marine sediments (temperature-related proxy), and fluctuation of the mineral grain size (discharge-related proxy) is also controlled by the global climatic change (Figs. 3.7 and 3.8). Probably, sediment discharge due to ice melting played an important role in Lake Baikal because glaciers in the catchments (from the Khamar-Daban Mountains in the south, the Baikalsky Mountains in the northwest, the Baruguzinsky Mountains in the northeast, and North Baikal Upland in the north) contributed large part of discharge into the lake during glacial periods,

K. Kashiwaya, *Geomorphology of Lake-Catchment Systems*,
Environmental Earth Sciences, DOI 10.1007/978-981-10-5110-4_8

although the amount of discharge was much smaller than during interglacial periods (Osipov and Khlystov 2010). It is estimated that the lake water level was comparatively constant between glacials and interglacials and the decrease in the level was around 40 m, only during the early part of glacials (Osipov and Khlystov 2010) (very small compared to the maximum depth of 1637 m).

Glaciers in the catchment of Lake Khuvsgul (Hövsgöl) (northern part of the catchment; the East Sayan Mountain Range) also affected significantly the hydrogeomorphological conditions in the system. During glacial periods, sedimentary records from Lake Khuvsgul responded differently to long-term climatic changes from Lake Baikal; there were large water level fluctuations between glacial and interglacial periods and no outflow from the lake during glacial periods (Fig. 3.12). Probably, this was mainly related to arid conditions (annual precipitation was estimated at around 100 mm in the previous glacial period, whereas it is around 300 mm today, Fedotov et al. 2004) and related to the limited feeding area (catchment). HCl-soluble material (mainly $CaCO_3$), corresponding to water level change, is a good proxy for global climatic change during the past 250 kyr. Large grain size (sediment discharge) was detected not only in warm periods (high water level) but also in cold periods (low water level) in lacustrine sediments.

The present water depth in the deepest point of Lake Khuvsgul is approximately 250 m, and the water level change between glacials and interglacials was estimated more than 100 m, which is more than 1/3 of the maximum depth. This kind of large level change is detected neither in Lake Biwa nor in Lake Baikal. The sensitivity of lacustrine sediments to catchment conditions may be expressed by the catchment–lake (C-L) ratio (Dearing and Foster 1993). Some conditions for main cores obtained in the three sites are shown in Table 8.1, indicating that the C-L ratio for Lake Khuvsgul is the smallest among the three. Therefore, water level change, which is linked to the size of the erodible zone of the naked bottom floor, could also be an important factor to consider when establishing a model for the Lake Khuvsgul system.

Table 8.1 Sampling conditions and C-L ratio for Lake Biwa, Lake Baikal, and Lake Khuvsgul

	Lake Biwa	Lake Baikal (BDP-98)	Lake Khuvsgul (HDP-04)
Sampling year	1983	1998	2004
Core length	1400 m	600 m	81 m
Water depth	68 m	333 m	250 m
Sampling point	35°13′064″ N, 136° 00′49″ E	53°44′48″ N, 108°24′ 34″ E	50°57′19″ N, 100° 21′32″ E
Level of water surface	84 m (alt.)	456 m (alt.)	1645 m (alt.)
C-L ratio	5.74 (3,174 km^2: 672.9 km^2)	17.14 (539,807 km^2: 31,494 km^2)	1.78 (4,913 km^2: 2,760 km^2)

8.2 Phenomenological Models on Lake-Catchment Systems for Long-Term Environmental Fluctuations

In a lake-catchment system, some materials produced due to external force (melting and/or rainfall) and internal forces (tectonic activity) in the catchment are transported to the lake and others are deposited in the catchment. Fallout substances in the catchment are also included in the deposited materials. The sedimentation rate per unit area for a lake is related to the size of the lake-catchment ratio if changes in other conditions are not so large. The grain size of transported materials is related to flow velocity; most coarse particles transported out of catchments generally deposited in the near river mouths, whereas fine particles transported with the lake current are deposited farther on the lake bottom (Ochiai and Kashiwaya 2003). Therefore, the determined grain size also depends on the depositional sites, which is also controlled by the water level change.

There are some differences in particle mobility conditions between the Lake Baikal system and the Lake Khuvsgul system. Comparatively, large materials (mineral grain size) were deposited during warm periods, suggesting high water discharge (rainfall and/or melt water) during warm periods in Lake Baikal, whereas the mineral grain size fluctuation in Lake Khuvsgul is not simple to interpret. Increases in the grain size during low water periods (glacials and/or stadials; aridity intervals) are supposed to indicate comparatively large water discharge (rainfall, melting, etc.). Pulse-like increases (including turbidites) in the boundaries between glacials (and/or stadials) and interglacials (and/or interstadials) are possibly related to the rapid enlargement of the surface area (exposed lake floor area), which is eroded in the beginning of glacials and/or stadials and rapid ice melting and precipitation increase at the onset of interglacials and/or interstadials.

Formulating a mathematical model is an important step for quantitatively understanding long lake-catchment processes and changes and for defining clear causal relations between the different factors. Furthermore, they are indispensable for future predictions. The model introduced here deals with the lake-catchment system on the assumption that lake sediments contain information on the environmental processes and changes in the catchments including channels. Here, the sedimentation rate in a lake is an important factor for establishing models, because the rate is directly linked to erosion (rate) and transportation (rate) in its catchment and in the lake.

Initial models of temporal environmental changes are generally phenomenological because the processes in these systems are complex. Establishing causal relations in the early stage is difficult. However, they are of great use for identifying factors to be quantitatively considered in complex systems such as lake-catchment systems. Here, a model of a comparatively long-term and relatively large-scale lake-catchment system is proposed. This kind of model is expected to clarify both phenomena of regional and global environmental changes. A figure prompting the model discussed here is shown in Fig. 3.7, which indicates fluctuations in the bi-SiO_2 concentration (nearly equivalent to the whole grain size) and the mineral grain size of Lake Baikal during the Brunhes epoch (Ochiai and Kashiwaya 2005).

The bi-SiO_2 concentration is a good proxy for global climatic (temperature) changes in Lake Baikal (Williams et al. 1997). Mineral grain size fluctuations, however, seem to be influenced more by regional conditions and are mainly related to the fluvial environment (sediment discharge). The fluvial environment itself has been closely related to glacier advance and retreat. According to a numerical simulation for a conceptual Lake Baikal, grain size fluctuations correspond to sediment discharge inflow from outside the lake (catchment) (Ochiai and Kashiwaya 2003). Then, a model of sediment discharge (erosion or mass production, mass transport) in the catchment and sedimentation in the lake is introduced (initial idea was given by Kashiwaya 2011). The ideal plane figure and the longitudinal sectional figure are shown in Fig. 8.1. In this case, the unerodible part comprises permafrost, base rocks, and forestry zones. Water level change in the lake is limited. Thus, a simplified model of the lake-catchment system introduced in the previous section is used to clarify the relation between erosion/transportation (mass production, mass transport, mobility) in the catchment and sedimentation in the lake. It is also assumed that the sedimentation rate in the lake ($SR(t)$) is proportional to the rate of erosion (production) in its catchment ($\mathrm{d}M(t)/\mathrm{d}t$), which yields the same equation as in the previous section:

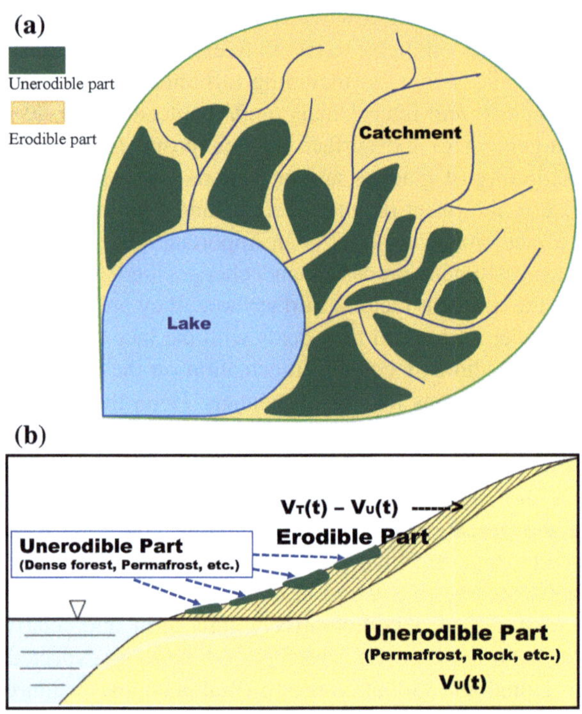

Fig. 8.1 **a** Plane figure of an ideal lake-catchment system. **b** Cross section of an ideal lake-catchment system. Lake; *light blue*, erodible part, *hatched ocher*; unerodible part, *green* and *ocher*; river, *dark blue*

$$\frac{\mathrm{d}S}{\mathrm{d}t}(t) = \alpha \frac{\mathrm{d}M(t)}{\mathrm{d}t}, \tag{8.1}$$

where $M(t)$ is the volume of catchment material eroded (transported) at time t, and α is a proportionality factor (sediment delivery factor including catchment structure properties). The volume of erodible materials at time t in the catchment ($V(t)$) is given by:

$$V(t) = \{V_T(t) - V_U(t)\} - M(t), \tag{8.2}$$

where $V_T(t) - V_U(t)$ corresponds to the volume of materials to be eroded in the catchment ($=V_E(t)$). $V_T(t)$ is the total volume of materials in the catchment, and $V_U(t)$ is the unerodible part of the materials at time t. Then, the erosion (production) rate is proportional to the size of the erodible part of the catchment ($V(t)$) at time t:

$$\mathrm{d}M(t)/\mathrm{d}t = \lambda(t)V(t), \tag{8.3}$$

where $\lambda(t)$ is an external factor relating to erosion (relative erosional force and climatic factor). Substituting (8.2) into (8.3), we obtain:

$$\frac{\mathrm{d}M(t)}{\mathrm{d}t} = \lambda(t)\{V_E(t) - M(t)\} \quad V_E(t) - M(t) \geq 0. \tag{8.4}$$

The equation itself is simple and easily solved. Assuming that $M(0) = M_0$ for the initial condition ($t = 0$), we obtain:

$$M(t) = \exp\left[-\int_0^t \lambda(\tau)\mathrm{d}\tau\right] \bullet \left[\int_0^t \{\lambda(\tau)V_E(\tau)\}\left\{\exp\int_0^\tau \lambda(\kappa)\mathrm{d}\kappa\right\}\mathrm{d}\tau + M_0\right]. \tag{8.5}$$

We obtain the following expression for the sedimentation rate (relative sediment discharge):

$$\begin{aligned} SR(t) &= \alpha\frac{\mathrm{d}M(t)}{\mathrm{d}t} \\ &= -a\lambda(t)\left[\exp\left[-\int_0^t \lambda(\tau)\mathrm{d}\tau\right] \bullet \left[\int_0^t \{\lambda(\tau)V_E(\tau)\}\left\{\exp\int_0^\tau \lambda(\kappa)\mathrm{d}\kappa\right\}\mathrm{d}\tau + M_0\right] - V_E(t)\right]. \end{aligned} \tag{8.6}$$

It is necessary to introduce some functions of the part to be eroded and external force for obtaining the sedimentation rate with time. The part to be eroded is expressed by:

$$V_E(t) = V_T(t) - V_U(t), \tag{8.7}$$

where $V_T(t)$ is related to tectonic activity ($\delta(t)$-uplift, subsidence, etc.), and it is a function of the activity $f_T(\delta(t))$. $V_U(t)$ includes the permafrost zone and dense forestry area in addition to the base rock part, which prevents surface materials from erosion. We assume that most of this part is a function of the climatic conditions ($f_{CW}(\lambda(t))$). Change in the base rock volume is considered a function of weathering (also related to climatic conditions) and tectonic activity (uplift, subsidence, etc.), i.e., $f_U(\lambda(t), \delta(t))$.

A large change in the exposed floor zone related to glacial–interglacial alternations such as those in the Lake Khuvsgul (Fig. 8.2a) is another important factor determining erosion and sedimentation. The exposed zone in glacial or low water periods is easily eroded due to external forces (rainfall, wind, etc.), and it is expressed as a function of the climatic conditions, but is inversely correlated to temperature ($f_{CC}(-\lambda(t))$); the effect of climatic conditions may be amplified in this setting. These lead to:

$$\begin{aligned} V_E(t) &= F_E\{f_T(\delta(t)), f_{CW}(\lambda(t)), f_{CC}(-\lambda(t)), f_U(\lambda(t), \delta(t))\} \\ &= F_{VE}\{\lambda(t), \delta(t)\} \geq M(t). \end{aligned} \tag{8.8}$$

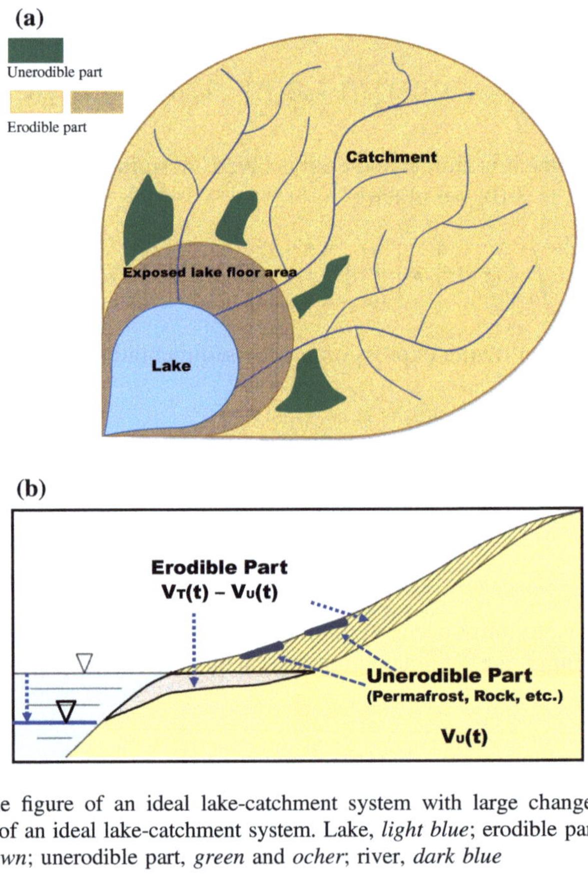

Fig. 8.2 **a** Plane figure of an ideal lake-catchment system with large change in water level. **b** Cross section of an ideal lake-catchment system. Lake, *light blue*; erodible part, *hatched ocher* and *hatched brown*; unerodible part, *green* and *ocher*; river, *dark blue*

In the simple case that the function is only related to climatic conditions (tectonic activity is limited), it may be assumed that:

$$V_E(t) = f_{VE}(\lambda(t)) = k\lambda(t) + m \geq M(t), \tag{8.9}$$

where k and m are coefficients.

Then, we obtain:

$$SR(t) = -\alpha\lambda(t)\left[\exp\left(-\int_0^t \lambda(\tau)\mathrm{d}\tau\right) \bullet \left[\int_0^t \{k\lambda^2(\tau) + m\lambda(\tau)\}\left\{\exp \int_0^\tau \lambda(\kappa)\mathrm{d}\kappa\right\}\mathrm{d}\tau + M_0\right] - k\lambda(t) - m\right]. \tag{8.10}$$

Now, let us discuss some hypothetical cases, assuming a simple oscillation of relative external force, to understand better the basic processes of the model;

$$\lambda(t) = p \sin qt + r. \tag{8.11}$$

Two cases are discussed. In the first case, there is a smaller change in the erodible area of the lake floor (large catchment-to-lake ratio and/or limited exposed floor zone) such as that in Lake Baikal (Fig. 8.1). In the second case, there is a larger change in the erodible area between glacial and interglacial periods (low water level and high water level), such as in Lake Khuvsgul (Fig. 8.2).

Some numerical values are introduced for the two cases as image experiments for the proper interpretation of the fluctuations in the relative sedimentation rate (Eq. (8.10)); this is related to a small effect from water level change (corresponding to Lake Baikal system). Some parts of the calculated results are shown in Fig. 8.3, assuming that $\alpha = 1$, $k = 0.05$, $m = 0$, $V_O = 1$, $p = 0.5$, $q = 40$, and $r = 1.0$ in Eqs. (8.9) and (8.11) for a numerical simulation. There are some patterns in the sedimentation rate responding to the climatic oscillation: The rate responds to a comparatively early stage of one oscillation cycle in some cases, while it responds to a comparatively late stage in other cases. In some cases, two peaks of the sedimentation rate appear in one climatic cycle. All results show that the relative sedimentation rate fundamentally follows external forcing (global environmental (climatic) change), but does not always respond linearly to the change; it is also controlled by the local/regional conditions. This is supported by the two observed fluctuations in the bi-SiO_2 content and in the mineral grain size for the Baikal system, as shown in Fig. 3.7. Small oscillations in mineral grain size (blue thin line) in the Baikal dataset (BDP98) may be related to global fluctuations (glacial change) with local catchment conditions, whereas large fluctuations (thick line, change in bi-SiO_2 content) may have been controlled by orbit-related fluctuations. A similar relation between the bi-SiO_2 content and the normalized mineral grain size is roughly detected in the Lake Biwa system although it is not so clear (Fig. 3.3). This may be mainly related to the absence of a glacier in the catchment, where temperature-related proxy has been directly linked to global climatic changes. Local conditions not directly linked to the changes may have been more significant for non-glaciated systems.

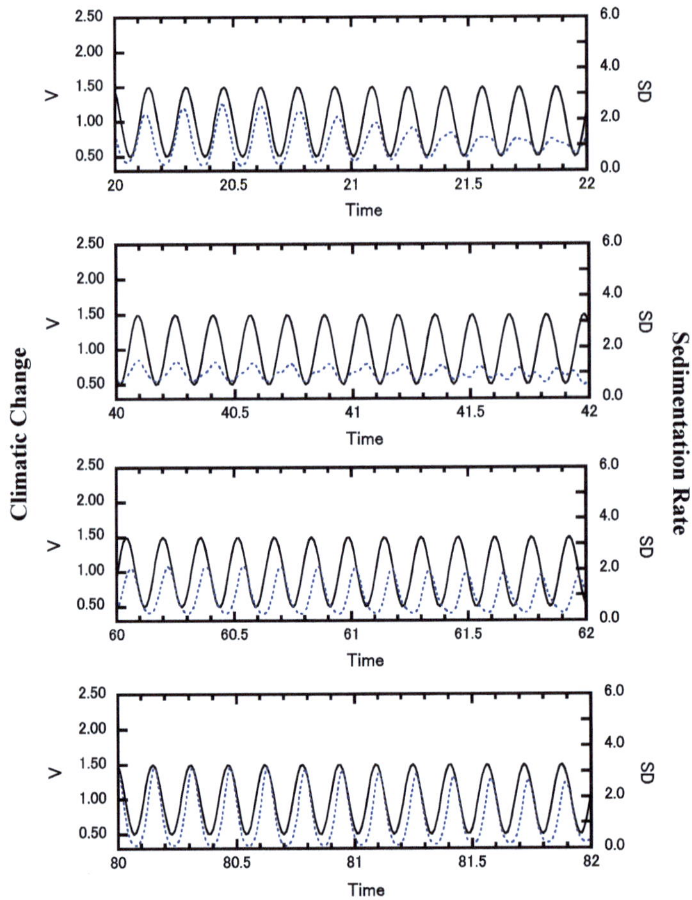

Fig. 8.3 Calculated results for an ideal lake-catchment system with small water level change. Imaginary climatic change, *solid black line*; sedimentation rate, *dotted blue line*

The model with some assumptions introduced here is plausible (even in the simple case that the volume to be eroded is a function of only climatic change) for explaining the mineral grain size fluctuations (local conditions linked to global changes) and the bi-SiO_2 content fluctuation (global change) in the first step.

Next, let us discuss a case where there were large changes in the erodible zone between glacial and interglacial periods (low and high water levels) such as in Lake Khuvsgul. The water level change is related to the size of the erodible part; low water level extends the erodible area in the lake floor during glacial periods. The unerodible forestry zone was also shrunk during these periods. Then, much larger values may be assumed for the proportionality factors; $k = 0.9$ and $m = 0.1$ (others are the same) are used in Eq. (8.10) for a numerical simulation. Some intervals of the calculated results are shown in Fig. 8.4, indicating that the response of the

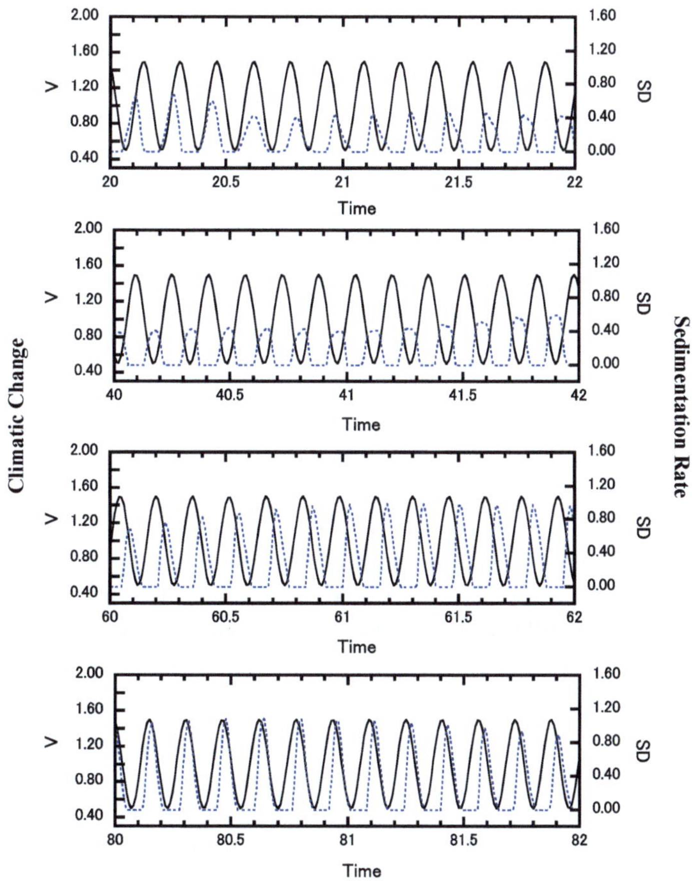

Fig. 8.4 Calculated results for an ideal lake-catchment system with large water level change. Imaginary climatic change, *solid black line*; sedimentation rate, *dotted blue line*

sedimentation rate to external forcing (climatic oscillation) is complex, as compared to the case of limited water level change, as shown above. Four types of responses to external forcing are detected: (1) The rate basically follows the climatic oscillation (it is large during the large external force (high-temperature stage) in the oscillation); (2) it shows maximum values before the maximum peaks in the oscillation; (3) it shows maximum values after the maximum peaks in the oscillation; and (4) it reaches its maximum values in the small external force (low-temperature stage) (the rate inversely responds to the climatic oscillation). All these patterns can be found in the fluctuations of the $CaCO_3$ content (a proxy of climatic oscillation) and the mineral grain size (a proxy for the sedimentation rate) for Lake Khuvsgul (Fig. 3.11). This suggests that the sedimentation rate may have responded to global environmental (climatic) changes, following various local conditions. There are some reasons for the various responses in Lake Khuvsgul; the lake-catchment ratio is very large (1.78), which means that the sedimentary environment in the lake is very sensitive to changes in the catchment conditions. Large water level changes were detected between glacials and interglacials (Krivonogov et al. 2005) (Fig. 3.12). Elongation of rivers and expansion of the erodible area due to water level drop prompts large changes in erosion and sediment transportability. These lead to that the model introduced here (a basic mathematical model expressed with Eq. (8.10) with some numerical values) is plausible for explaining the relation between $CaCO_3$ (a proxy for climate) and mineral grain size (a proxy for sediment discharge) in Lake Khuvsgul.

The discussion introduced here is the first step for a proper understanding of causal relations in lake-catchment systems based on lake sediment information. Models are established based on simplified imaginary ideas, and they are phenomenological ones in the present stage. Geomorphic systems expressing long-term earth-surface processes and topographical changes are too complex to be explained with simple models. However, long-term data relating to such systems may be physically explained to a certain extent if proper mathematical models are introduced for the systems. Numerical calculations based on the mathematical model introduced here show that the difference in the response to climatic changes (external force) between Lake Baikal (Lake Biwa) and Lake Khuvsgul (sedimentary fluctuations) may be roughly explained by the model, supporting that the model is a plausible one as a first approximation for the next step to interpret more essentially long-term changes in lake-catchment systems.

References

Dearing JA, Foster IDL (1993) Lake sediments and geomorphological processes: some thoughts. In: McManus J, Duck RW (eds) Geomorphology and sedimentology of lakes and reservoirs. Wiley, Chichester, 278p, pp 5–14

Fedotov AP, Chebykin EP, Yu Semenov M, Vorobyova SS, Yu Osipov E, Golobokova LP, Pogodaeva TV, Zheleznyakova TO, Grachev MA, Tomurhuu D, T Oyunchimeg, T Narantsetseg, Tomurtogoo O, Dolgikh PT, Arsenyuk MI, De Batist M (2004) Changes in the volume and salinity of Lake Khubsugul (Mongolia) in response to global climate changes in the upper Pleistocene and the Holocene. Palaeogeogr Palaeoclimatol Palaeoecol 209:245–257

Kashiwaya K (2011) Lake-catchmen tsystems and sediment information in Baikal district (Siberia and Mongolia). J Earth Environ 2:417–425

Krivonogov SK, Sheinkman VS, Mistruykov AA (2005) Stages in the development of the Darhad dammed lake (Northern Mongolia) during the late Pleistocene and Holocene. Quat Int 136:83–94

Ochiai S, Kashiwaya K (2003) Hydro-geomorphological changes and sedimentation processes printed in sediments from Lake Baikal, In: Kashiwaya K (ed) Long Continental Records from Lake Baikal. Springer, Tokyo, 370p, pp 297–312

Ochiai S, Kashiwaya K (2005) Climato-hydrological environment inferred from Lake Baikal sediments based on an automatic orbitally tuned age model. J Paleolimnol 33:303–311

Osipov EY, Khlystov OM (2010) Glaciers and meltwater flux to Lake Baikal during the Last Glacial Maximum. Palaeogeogr Palaeoclimatol Palaeoecol 294:4–15

Williams DF, Peck J, Karabanov EB, Prokopenko AA, Kravchinsky V, King J, Kuzmin MI (1997) Lake Baikal record of continental climate response to orbital insolation during the past 5 million years. Science 278:1114–1117

Chapter 9
Long-Term External Forcing and Limnogeomorphology

9.1 External Forcing and Solar Activity

There are two forces that have controlled earth-surface processes and changes: internal (endogenic) forces and external (exogenic) forces. Here, only the external forces are discussed. External forces are essential also in limnogeomorphology as well as in geomorphology in general. As shown in the previous chapters, it is indispensable to obtain appropriate information on external forcing through limnogeomorphological studies. There are two activities (natural and anthropogenic) concerned with these forces; however, only natural activity is a target for discussion for evaluating long-term effects. It is obvious that it is closely related to atmospheric conditions. Changes in natural processes are linked to climatic changes, and they present temporal variability. In previously introduced models, the term of $\beta(t)$ was used as a simplified form of the changes although they are complex.

One of the primary external forces on the earth surface is solar activity. In the short time scale, direct solar activity may be more related to external forcing, whereas solar radiation (insolation) is essential in the long time scale (e.g., the Cenozoic). Recently, in discussions on the effect of solar activity on climate, galaxy cosmic ray was addressed (there is a trade-off between solar activity and intensity of galaxy cosmic ray); changes in the intensity of galactic cosmic ray prompt cloud development in the lower troposphere (Fig. 9.1; Svensmark 2007, 2015; Svensmark and Calder 2007; Svensmark et al. 2012). Cloud development may be liked to precipitation increase. Reconstruction of solar activity over the past millennia was achieved using cosmogenic isotopes (^{14}C) in dated tree rings (Fig. 9.2; Usoskin et al. 2007, 2008). The intensity of galactic cosmic ray, as indicated through the ^{14}C concentration, is closely related to changes in the $\delta^{18}O$ concentration of speleothem in the Alps (Fig. 9.3; Kirkby 2007). Advance and retreat of the glacier in the Andes was also corresponded to the intensity of galactic cosmic rays (intensity of ^{14}C and ^{10}Be; Polissar et al. 2006; Kirkby 2008) (Fig. 9.4). However, the mechanism for cloud development has been debated (e.g., Svensmark et al. 2012, 2013; Laken

K. Kashiwaya, *Geomorphology of Lake-Catchment Systems*,
Environmental Earth Sciences, DOI 10.1007/978-981-10-5110-4_9

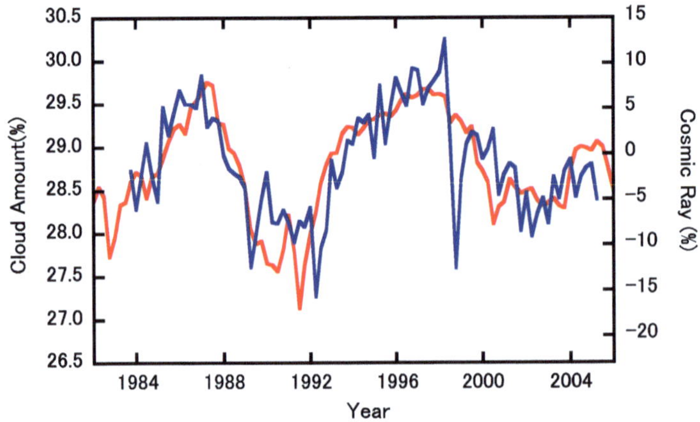

Fig. 9.1 A relationship between low cloud amount (*red*; %) and cosmic ray (*blue*, %) (modified after Svensmark 2007)

et al. 2012a, b; Dunne et al. 2012; Lockwood 2012; Krahenbuhl 2015). The idea of a relation between galaxy cosmic ray and climate has been fascinating not only for short-term changes but also for long-term changes, although further discussions on the mechanism are required.

In contrast, there have been numerous studies on the relation between solar insolation and climate (astronomical theory of climatic changes) before it was recognized as a basic theory of long-term climatic change in the Quaternary. Today, the astronomical theory of climatic changes is called Milankovitch theory after a theoretical climatologist who completed it. Brief introductions of the history are given by some books (Imbrie and Imbrie 1979; EGS 1995). There were three main scientists who contributed to establish the theory. The French mathematician J.A. Adhémar developed his idea in the book "Révolution des Mers" published in 1842 (Adhémar 1842). In this book, he proposed that the precession of the equinoxes controlled global climatic change; the precession has a cycle of approximately 21,000 years, which is a combination of the axial precession (clockwise rotation with a period of about 26,000 years) and the rotation of the elliptical orbit itself (counter-clockwise rotation with a period of approximately 110,000 years) in the same plane. Hence, the cycle for the shifting of the equinoxes along the orbit is 21,000 years (Fig. 9.5). The winter solstice in the northern hemisphere is now located in the place where the earth was close to the sun near one end of the ellipse, and it was located where the earth was far away from the sun near the opposite end of the ellipse ca. 10,500 years ago. The earth was at the same position ca. 21,000 years ago. Thus, traveling distance in the seasons changes with time. Today, the summer season in the northern hemisphere is longer than the winter season, whereas in the southern hemisphere, the winter season is longer than the summer season is. Adhémar considered that the northern hemisphere is now in the interglacial and the southern hemisphere is in the glacial phase because there are huge ice sheets in the Antarctic. His idea was rejected later because the total amount of

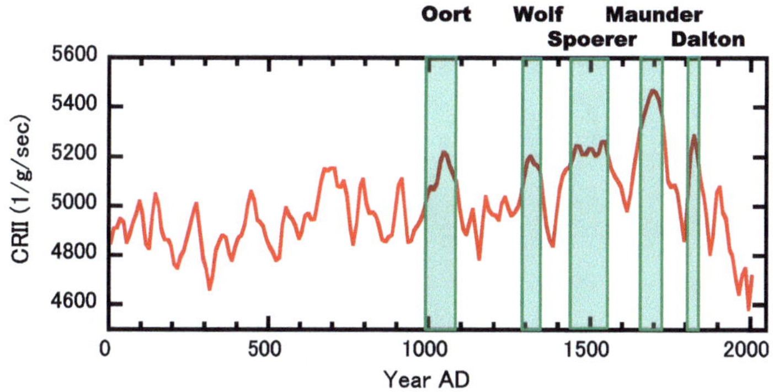

Fig. 9.2 Temporal change in cosmic ray induced ionization (CRII) averaged over some longitudes at 45° N and solar minimum events (*vertical green belts*) (modified after Usoskin et al. 2008)

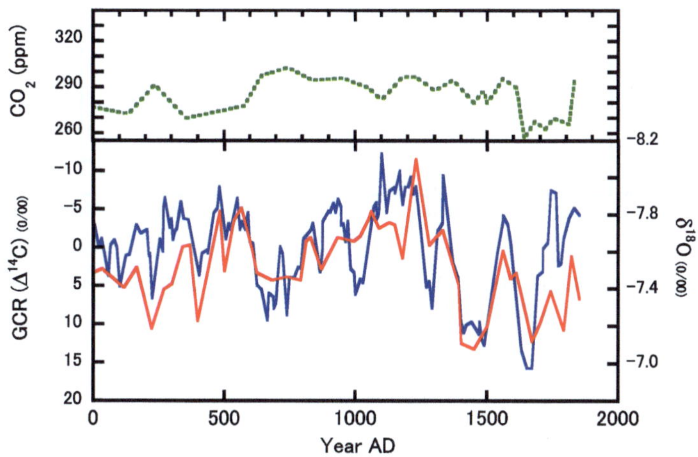

Fig. 9.3 Changes in galactic cosmic ray (blue line; GCR/^{14}C, ‰), δ^{18}O (*red line*, ‰) and CO_2 concentration (*green dotted line*, ppm) (modified after Kirkby 2008)

heat received by one hemisphere during the year is always the same as that received by the other. Nevertheless, the idea that the precession of the equinoxes plays an important role in global climatic changes was the first important step for establishing the astronomical theory of climatic changes.

The second runner in the history of the astronomical theory was James Croll (a Scottish geologist), who noticed that change in the eccentricity of the earth's orbit, which Adhémar had not considered, might be an important astronomical factor (Croll 1875). He showed that the intensity of radiation received by the earth during each season is affected by changes in eccentricity. If the orbit was nearly circular, the precession of the equinoxes would have little effect on climatic changes, for each season would then be located at nearly the same distance from the sun. This suggests that during low eccentricity periods, winters are not cold enough to establish glacials,

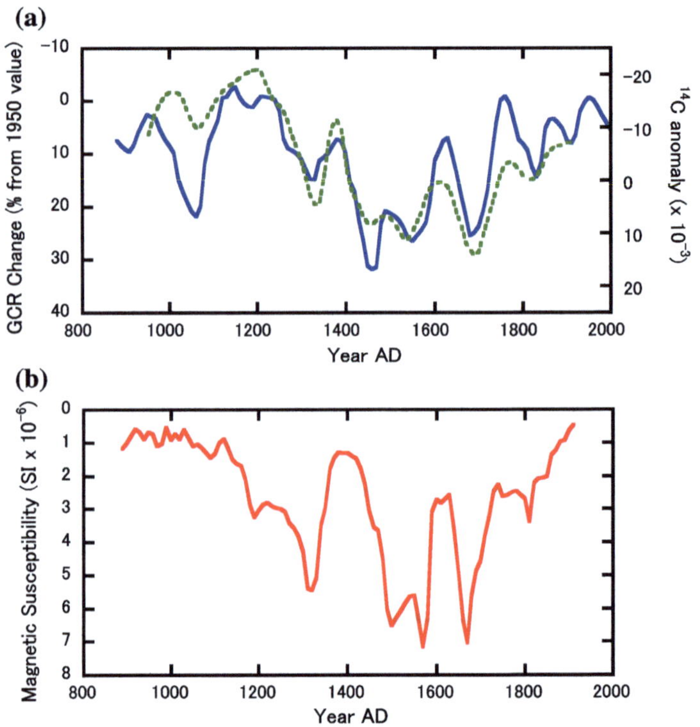

Fig. 9.4 **a** Changes in GCR (*blue*: ^{10}Be, *dotted green:* ^{14}C) and **b** magnetic susceptibility in Lake Mucubaji (modified after Kirkby 2008)

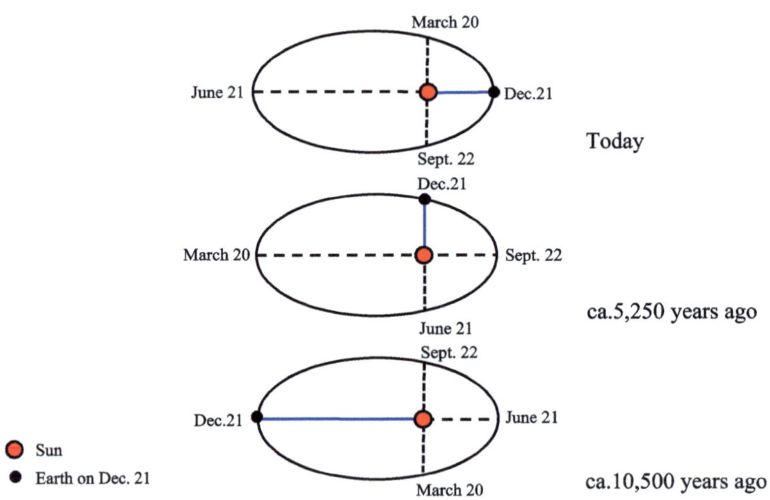

Fig. 9.5 Precession of the equinoxes; today to ca. 10500 years ago

irrespective of wherever the winter solstice is located on the earth's orbital path. However, during greater eccentricity periods, winters are warmer when the solstice is located near the sun at the short end of the orbit, whereas winters are much colder when it is located far from the sun at the long end of the orbital path. Croll believed that glacials corresponded to the distance between the Earth and the Sun, as measured on December 21 (winter solstice) (Imbrie and Imbrie 1979). When this distance exceeds a critical value, winters are cold enough to cause glacials in the northern hemisphere. On the contrary, in the southern hemisphere, glacials occur when the distance is less than a critical value. Croll thought that the long intervals during which the eccentricity of the orbit is large enough to cause ice periods in one hemisphere or the other should be called glacials (glacial periods), and the intervals in between should be called interglacials (interglacial periods). He considered that the last glacial (period) began approximately 250,000 years ago and ended approximately 80,000 years ago; the Earth has been in an interglacial (period) since then. This was not accepted because it was considered then that the end of the last glacial had been much younger. Additionally, climatic changes due to solar heating based on his idea were questioned because solar insolation change was considered too small to produce such changes. However, Croll noticed Le Verrier's idea of temporal change in the tilt of the Earth's axis (ca. 22°–25°), which would also cause glacial initiation.

The idea was thoroughly discussed and developed by Milutin Milanković (Milankovitch) (a Serbian geophysicist), an anchor of the astronomical theory of climate. In fact, there is an interesting scientific story on his idea's application to studies on climatic changes, geomorphological phenomena, etc. The starting point of the story was that the book he published (Milanković 1920) attracted the attention of Wladimir Köppen, a famous climatologist. In the book, the present climates on Earth, Mars, and Venus were mathematically described; orbital variations changing the geographic and seasonal distribution of sunlight were sufficient to produce ice periods; and it was possible to calculate the solar radiation on the Earth for any time in the past. Following Köppen's suggestion that the summer radiation in the middle latitude mostly determines the increase and decrease in the ice volume of the Earth, Milankovitch calculated values for summer radiation at latitudes 55°, 60°, and 65° North during the past 650,000 years. Geomorphological findings on the history of the Alpine glaciers were published by the German geographers Albrecht Penck and Eduard Brückner in 1909; four layered glacial materials (moraine and till) were corresponded to four glacial advances in the Alps during the Quaternary period (Penck and Brückner 1909). It was with the four advances printed in geomorphological phenomena that the Milankovitch theory (Milanković 1920, 1941) was accepted in the field of climatology, although Penck rejected the theory in the beginning (EGS 1995). Köppen and Wegener (1924) indicated that the insolation curve could be matched reasonably well with the growth of the Alpine glaciers. Four bundles of low insolation intervals estimated with the theory could be correlated to the four glacial periods (Günz, Mindel, Riss, and Würm) shown in the Alpine field (Fig. 9.6). This was one of the earliest evidences that orbital fluctuations were imprinted in the geomorphic development. It was probably the fact that the long interglacial interval between the Mindel and Riss glacial periods was identified in the

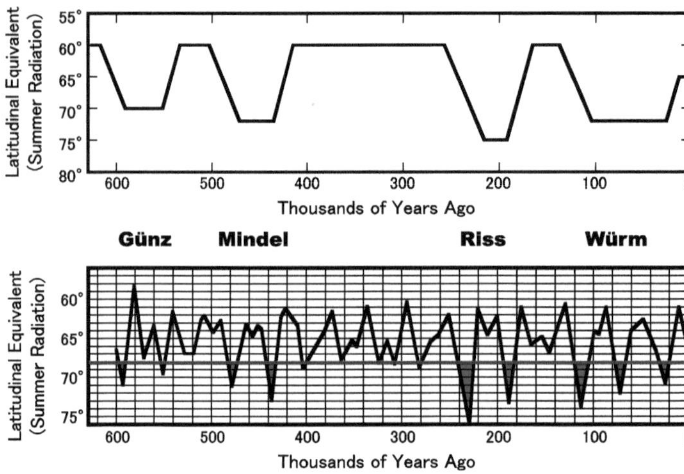

Fig. 9.6 Estimated glacial series in the Alps suggested by Penck and Brückner (1909) (*upper*) and insolation curve for latitude 65° North (calculated with the results by Le Verrier and Mishkovitch) (*lower*) (modified after Milanković 1941)

long warm interval of the insolation curve that convinced Köppen and Wegener that the insolation curve of Milankovitch corresponded to the glacier fluctuations. Much later, this also indicated that landform developments could be prompted by orbital fluctuations through environmental changes in lake-catchment systems. This idea is linked to limnogeomorphology.

Afterward, Milankovitch said, "my work based upon exact science passes into the sphere of the descriptive natural sciences and thereby forms a bridge between the exact and the descriptive natural sciences. It is the hitherto missing link between celestial mechanics and geology" (1941). The theory itself was once partially rejected because it was not supported with exact age models when the ^{14}C dating method was developed (Imbrie and Imbrie 1979; Imbrie et al. 1984), although some scientists thought that it was essentially not denied (e.g., Holmes 1965; Scheidegger 1970). Then, the theory was dramatically revived in the ocean as Wegener's theory on continental drift emerged (EGS 1995). Orbital cycles (one eccentricity, one obliquity, and two precession cycles) based on the Milankovitch theory were found in the $\delta^{18}O$ fluctuation of oceanic bottom sediments (Fig. 9.7; Hays et al. 1976).

Since then, the Milankovitch theory has obtained a dominant position in long-term climatic theory (especially, the Quaternary period) with revised calculations (e.g., Fig. 9.8) (Berger 1978; Berger and Loutre 1991; Laskar et al. 1993, 2004, 2011).

For example, insolation minima at about 3.9, 3.8, and 3.7 Ma with the revised calculation by Lasker et al. (1993), which were obliquity-related, may have triggered Milankovitch-type climatic oscillations (with periodic major Northern Hemisphere ice sheets) printed in the Lake Baikal sediments. Intensification of these oscillations at approximately 2.8 Ma was probably connected to another decrease in insolation (Kashiwaya et al. 2003a). An additional longer cycle around

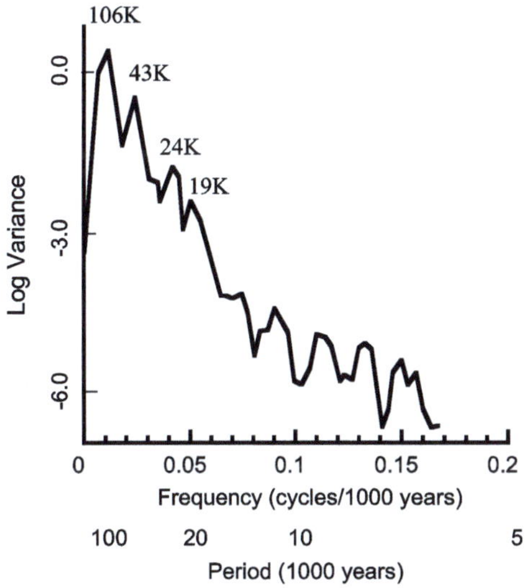

Fig. 9.7 Milankovitch cycles (106 K, eccentricity; 43 K, obliquity; 24 K and 19 K, precession) printed in oceanic sediments (modified after Hays et al. 1976)

2.4 Myr related to the eccentricity calculated by Laskar et al. (2011) may provide a clue for much longer climatic oscillation, which appears in the long Baikal sediments (Fig. 9.9) (Kashiwaya et al. 2003b).

Several issues remain open for discussion, such as the MIS-11 problem (super interglacial period around 400 ka) and the 100 kyr cycle problem. The MIS-11 problem has been discussed in the context of the present interglacial extension (another super interglacial), because the insolation fluctuation in the 30 kyr interval around present (past 10 kyr to the next 20 kyr) is similar to that in the 30 kyr interval around MIS-11 (Droxler et al. 2003). However, continental records from the middle latitudes (Lake Baikal, Lake Khuvsgul, and Lake El'gygytgyn) show a cold period in this interval corresponded to insolation, although some additional data are needed to develop more exact age models for the sedimentary records (Fig. 9.10) (Kashiwaya et al. 2010) (Karavanov et al. 2003; Ochiai and Kashiwaya

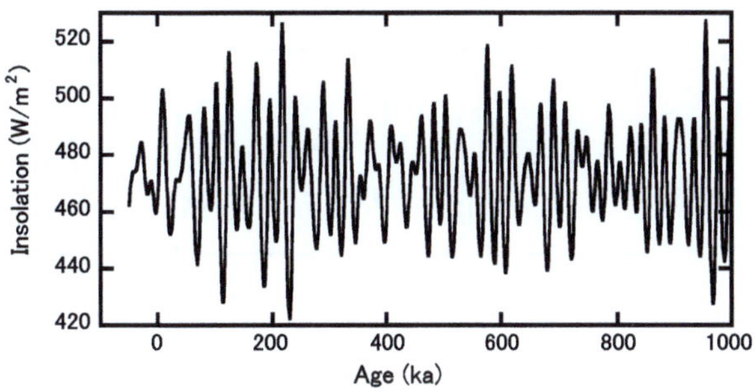

Fig. 9.8 Insolation curve for latitude 65° North during the past 1000 kyr and next 50 kyr (original data by Laskar et al. 1993)

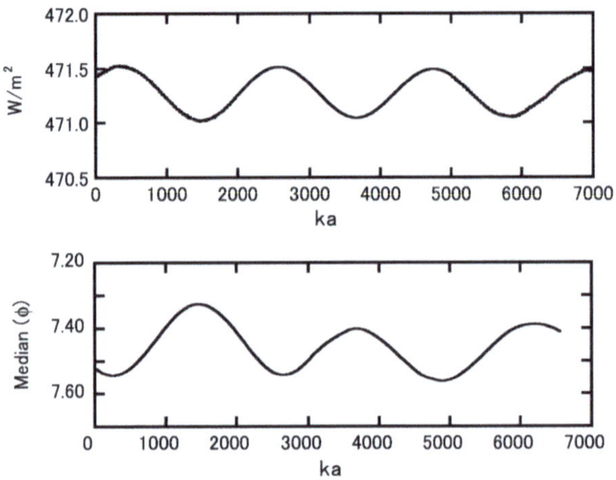

Fig. 9.9 Filtered insolation curve (1.5–2.5 my) (*upper*) and mineral grain size in Lake Baikal (*lower*) (modified after Kashiwaya et al. 2003a, b)

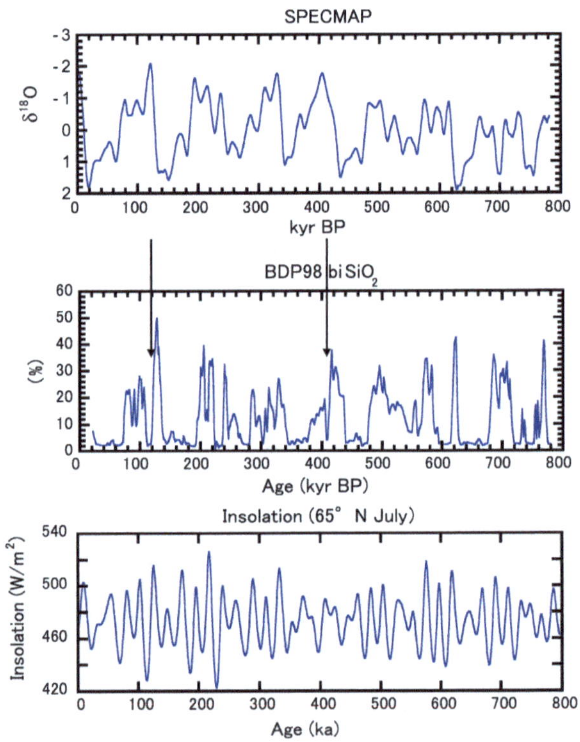

Fig. 9.10 Fluctuations in **a** oceanic data ($\delta^{18}O$) (SPECMAP Archive, Imbrie et al. 1989), **b** lacustrine data (bi-SiO_2 in Lake Baikal), and **c** insolation at 65° North (Laskar et al. 1993). Arrows indicate the cold periods at the MIS-5d and during the MIS-11

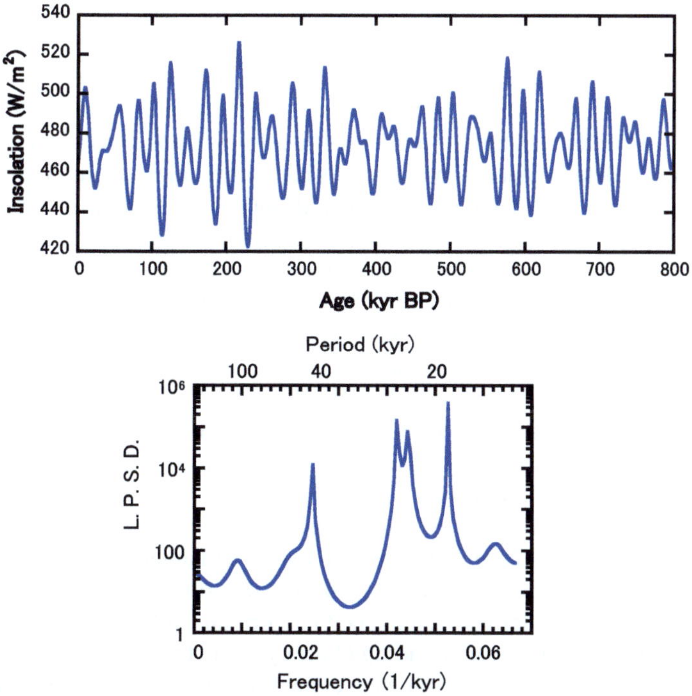

Fig. 9.11 Insolation curve in the mid-July at 65° North (*upper*; Laskar et al. 1993) and spectral analysis for the curve (*lower*)

2005; Vogel et al. 2013). Further discussion on the issue may be required. The 100 kyr problem has been a target of discussion since the Milankovitch theory was revived in 1976; why the small amplitudes around the 100 kyr-period in the insolation exhibit the largest amplitudes in the environmental records (Fig. 9.11). Many ideas have been reported; among them, ice sheet models have been developed as one of the most promising ideas (e.g., Weertman 1976; Pollard 1983; Abe-Ouchi et al. 2013).

9.2 Long-Term External Forces Recorded in Lacustrine Sediments

Pioneering works aimed at reconstructing long-term environmental changes in terrestrial areas were commenced in Lake Biwa, Japan, in the early 1970s (Horie 1973). Lacustrine sediments include high-resolution information on lake-catchment systems and their surrounding environment because of the comparatively large sedimentation rate. They can be used to understand terrestrial processes, especially lake-catchment processes, including physical processes in lakes and catchments.

External forces as well as internal forces play an important role in reconstructing past landforms and estimating future ones in geomorphology, as shown in the above chapters (e.g., Kashiwaya 1994, 2012). Furthermore, long-term earth-surface changes may be reconstructed using oceanic sediment information (filtered information), but an understanding of individual landform changes requires information that has greater detail. Glacial–interglacial scale information might be available for landform changes in the Quaternary period such as the four glacial advances (landform changes) found in the Alps mentioned above. However, the information is too rough for quantitatively reconstructing and predicting landform changes due to both natural and human activities. However, lacustrine sediment information could be available for various scale phenomena, depending on the local lake-catchment conditions, as shown above.

The Lake Biwa sedimentary records cover the last 4 Myr, but detailed information is limited in the upper part of the sediment (250 m; ca. 450 ka). The core (200 m) obtained in 1971 was the first longest academic lacustrine core ever drilled, but unfortunately, the age model for the core was not so exact in the beginning because it was mainly based on Fission truck dates (Horie 1984, 1987). Some reports indicated that Milankovitch cycles might be printed in the core (Yamamoto et al. 1984, 1985). However, it was after the longer core (1400 m) was obtained in 1983 that a comparatively exact age model was established and orbital cycles were clearly detected (Kashiwaya et al. 1991). The first finding of the cycles in the lacustrine sediments was in the fluctuation of the coarse grain size content (200-m core; Fig. 3.1), indicating that this factor could reflect some global climatic changes. The grain size fluctuation in the sediments does not always indicate some changes in hydrogeomorphological conditions (e.g., it may also be related to diatom size).

Samples from the 1400-m core were also analyzed for some items (Yamamoto 1995; Ishikawa 2004). The grain size fluctuation shown in Fig. 9.12 is a stacked curve comprised of the 200-m core and the upper clay part of the 1400-m core, which is available for establishing an age model with magnetic dates and orbital tuning. The detrended fluctuation based on the age model is shown in the lower with insolation (Fig. 9.12b). Changes in sediment composition (bi-SiO_2 and organic content; BIOR) (black) and mineral grain size (median; blue) are shown in Fig. 9.13. The black line (BIOR; temperature-related proxy) corresponds to global climatic change, and the fluctuation related to glacial and interglacial cycles are clearly detected in addition to the depression in the super interglacial around 400 ka (MIS11), as shown above.

Long-term changes in the mineral grain size show comparatively rapid decrease in the beginning and a gradual increase afterward. Mineral grain size is a proxy for hydrological conditions (precipitation, discharge, water level, etc.) as shown above. The long-term trend shown in this figure suggests some bathymetrical change (water level) in basin conditions; comparatively rapid bottom subsidence in the early stage (decrease in grain size) and gradual aggradation afterward (increase in grain size). It is estimated that the present deep water conditions in Lake Biwa were first established at approximately 250-m depth (ca. 450 ka) (Horie 1984). Changes in hydrological conditions may be more clearly detected to cancel the trend.

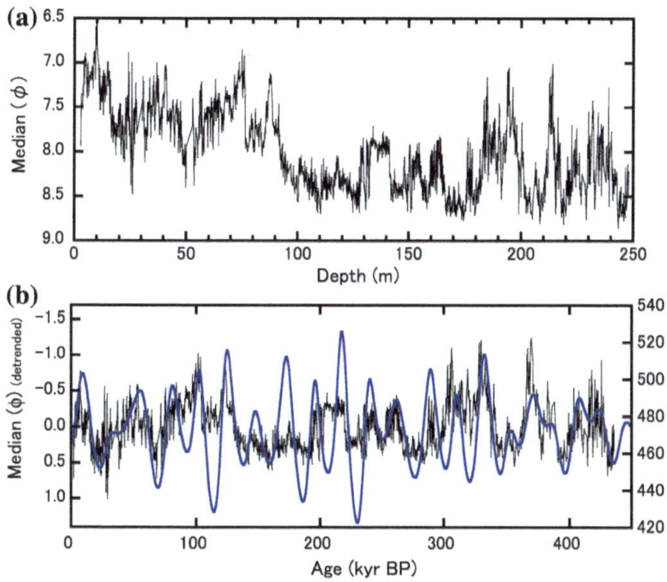

Fig. 9.12 **a** Grain size (median) fluctuations in a stacked dataset (200-m core and the upper clay part of 1400-m core) and **b** fluctuations in detrended grain size (*black*) and insolation (*blue*) in the past 450 kyr

Detrended BIOR and mineral grain size are shown in Fig. 9.14a, and they suggest that hydrological conditions were roughly controlled by global climatic (temperature) changes (solar insolation) but occasionally responded differently to the

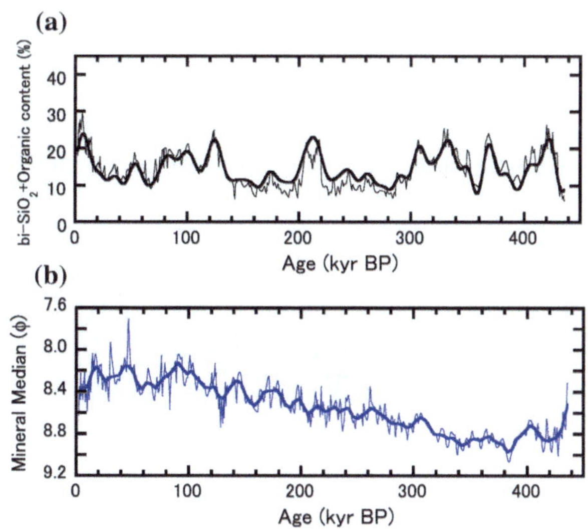

Fig. 9.13 **a** Changes in temperature-related proxy (bi-SiO_2 + organic content) (*black*), and **b** mineral grain size (*blue*). Filtered: *thick*, and original: *thin*

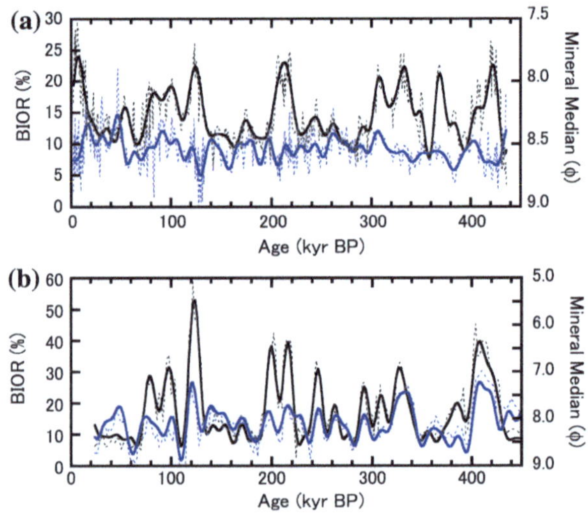

Fig. 9.14 Changes in detrended BIOR (bi-SiO_2 + organic content) (*black*: filtered, *solid;* original; *dotted*) and mineral grain size (*blue*: filtered, *solid;* original, *dotted*) in **a** Lake Biwa and in **b** Lake Baikal

changes (locally different response). This may be compared with the case of Lake Baikal (Figs. 9.14b and 9.16), indicating that most changes in both filtered mineral grain size records (thick lines; black from Lake Baikal, blue from Lake Biwa in Fig. 9.15) correspond positively to insolation (red line), and the original mineral grain size in Lake Baikal (thin black) has coarser values and larger amplitude than that of Lake Biwa (thin blue), except for the last 100 kyr.

There is a clear climato-environmental difference between the two areas; there is glacier advance in the catchment and iceberg production in the lake during glacials in the Baikal lake-catchment system (Eduard and Oleg 2010), but there is no glacier even during glacials in the Biwa system. Two kinds of proxy datasets (bi-SiO2 content + organic content; temperature-related proxies and mineral grain size (median); discharge-related proxies) are discussed in order to clarify the effect of

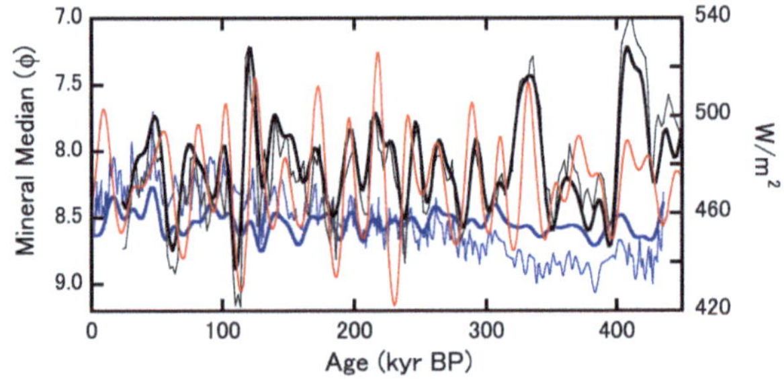

Fig. 9.15 Changes in mineral grain size in Lake Baikal (*black*: filtered, *thick;* original, *thin*), Lake Biwa (*blue*: detrended filtered; *thick*, original; *thin*), and insolation curve in the mid-July at 65° North (*red*)

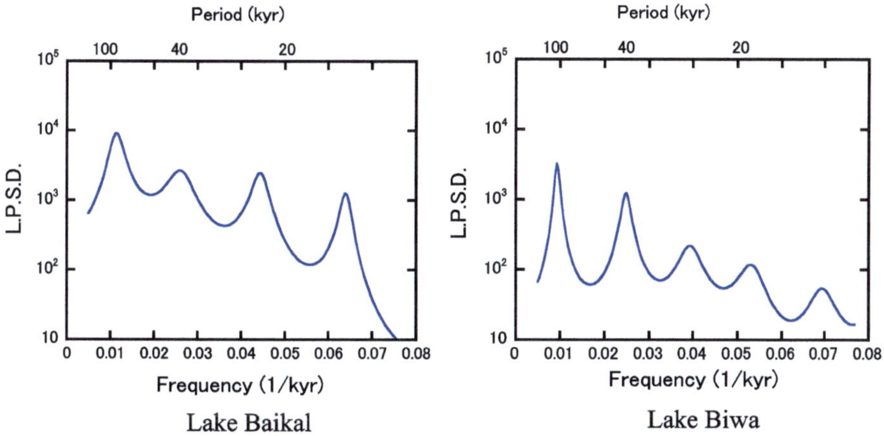

Fig. 9.16 Spectral analysis for the BIOR (bi-SiO_2 + organic content) in Lake Baikal (*left*) and Lake Biwa (*right*)

insolation on the earth-surface processes in the two areas. Calculated results with spectral analysis for the two proxy datasets of the two systems are shown in Figs. 9.16 and 9.17. As shown in the case of the coarse content in the Lake Biwa sediments (Fig. 3.1), the four Milankovitch cycles are clearly detected in the two proxy datasets (temperature-related proxies) of the both systems (Fig. 9.16). On the contrary, the cycles in the discharge-related proxy for the Lake Biwa dataset are not so marked (Fig. 9.17), although they are as clear as those in the temperature-related proxy in the Baikal record. Probably, this also reflects the environmental difference; remarkable increase in discharge, including glacier melting, is observed during interglacials in the Baikal system. The Lake Biwa system, located in climatically humid and tectonically active zone, presents complex responses to solar insolation in spite of basic solar control.

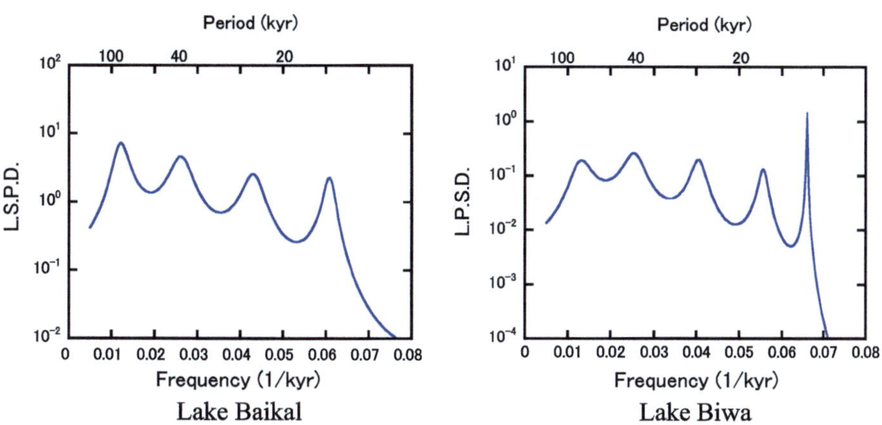

Fig. 9.17 Spectral analysis for the mineral grain size (median; ϕ) in Lake Baikal (*left*) and Lake Biwa (*right*)

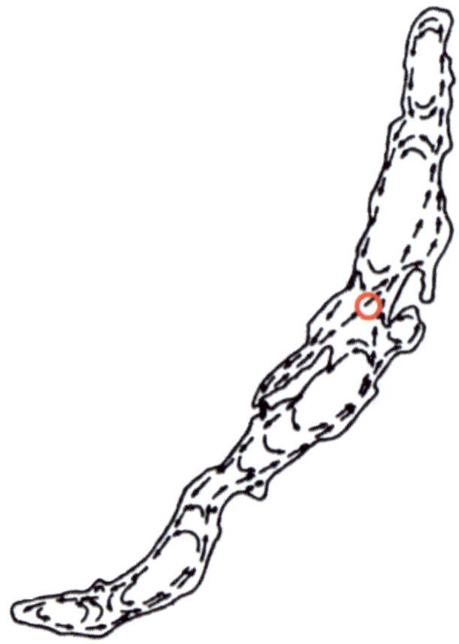

Fig. 9.18 Current direction in Lake Baikal (Shimaraev et al. 1994): 2–5 cm/sec at the depth of 10–15 m. *Red circle* means the sampling point of BDP98 (Academician Ridge)

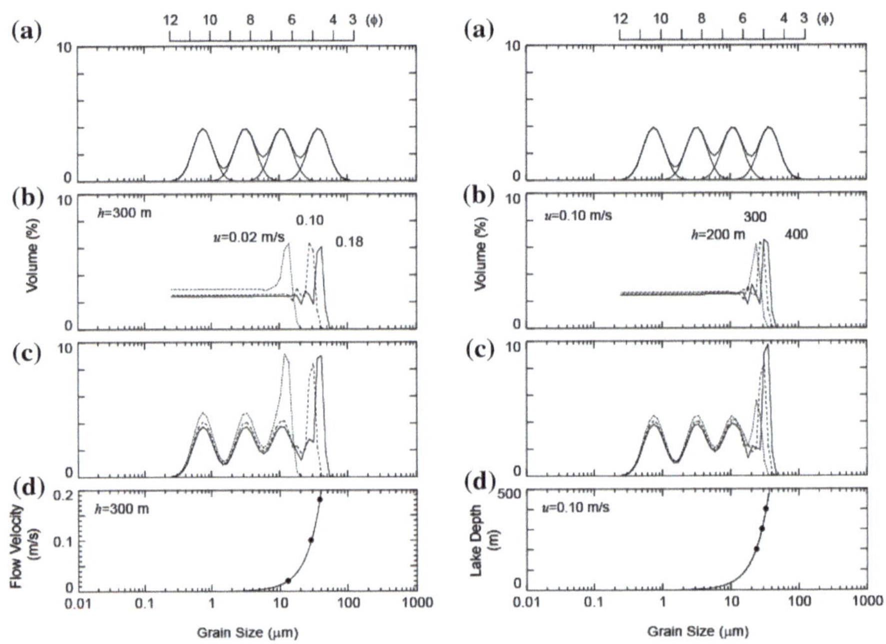

Fig. 9.19 Calculated grain size distribution (**a**, **b**, and **c**) and flow velocity (**d**; 0.02, 0.10, and 0.18 m/s) at the depth of 300 m (*left*), and grain size distribution (**a**, **b**, and **c**) and water depth (**d**; 200, 300, and 400 m) for the flow velocity of 0.1 m/sec (*right*) (Ochiai and Kashiwaya 2003)

As mentioned above, a temporal change in the original grain size (dotted blue) of Lake Biwa suggests some bathymetric change. The sampling point for the Biwa cores is located in the deeper part of the basin, at the depth of 70 m and 4 km from the west coast. The mouths of the nearest rivers are approximately 8–9 km from this point, which is now under the second gyre (clockwise, approximately 10 cm/s at the depth of 5 m) in the lake (Endoh and Okumura 1993). It is reported that the recent grain size distribution in the lake bottom may be controlled also by the gyres, depending on the bathymetric conditions (Kumon et al. 1993).

In contrast, the BDP98 sampling point is located on the Academician Ridge (water depth of 333 m) and 25–30 km from the west coast. In general, comparatively large particles cannot be transported to a convex area such as a ridge from the surrounding area without large sediment discharge such as turbidity currents. The present main currents above the Academician Ridge are those from the southwest and the south (Fig. 9.18), and the observed current velocity is 2–5 cm/s at the depth

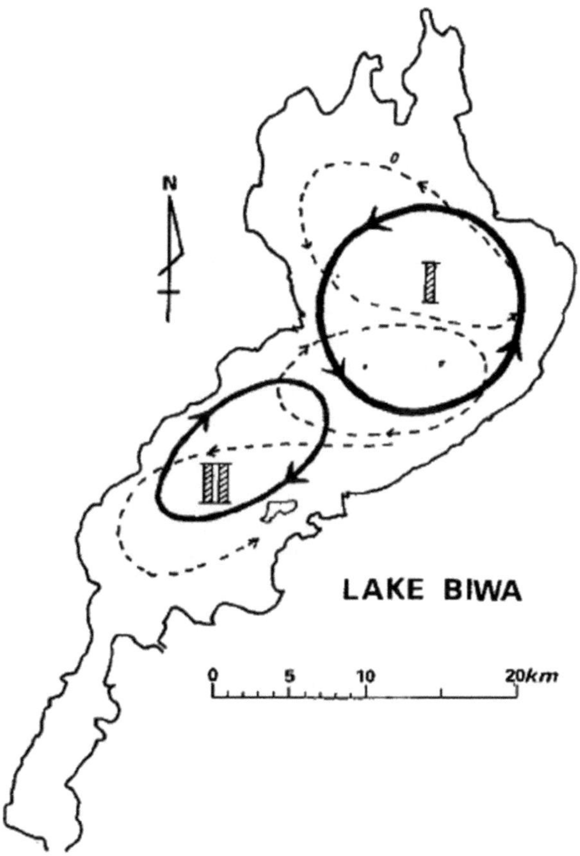

Fig. 9.20 Schematic pattern of the gyre system in Lake Biwa (*solid line*) and past estimated pattern (*dotted line*) by Suda et al. (1926). Current velocity, 10 cm/sec at the depth 5–7 m (Endoh and Okumura 1993)

of 10–15 m (Shimaraev et al. 1994). Ochiai and Kashiwaya (2003) remarked that the grain size distribution of the bottom sediment may mainly depend on the current velocity and additionally on the water depth if the riverine sediment input has a constant size distribution, based on a numerical simulation with a simplified conceptual model of Lake Baikal sedimentation (Fig. 9.19).

Turbidity (density) currents due to heavy rainfall, earthquakes, melting, etc., often affected both lake-catchment systems, probably closely related to the size of the sediment discharge through earth-surface processes (Taishi et al. 1991; Appleby et al. 1998; Shiki et al. 2000; Charlet et al. 2005). The frequency and size of the gyres and the density currents induced due to climatic and tectonic (seismic) activities may also affect the transporting potential in these systems (Endoh 1995; Endoh et al. 1995; Akitomo et al. 2009) (Fig. 9.20; Endoh and Okumura 1993), which requires more discussions on the physical processes of transportation of small grain sizes in the systems. Therefore, it is necessary to decipher the relations between the information on hydroenvironmental settings including earth-surface processes and changes, and the sedimentary conditions.

References

Abe-Ouchi A, Saito F, Kawamura K, Raymo ME, Okuno J, Takahashi K, Blatter H (2013) Insolation-driven 100,000-year glacial cycles and hysteresis of ice-sheet volume. Nature 500:190–193

Adhémar JA (1842) Révolutions de la Mer: Déluges Périodiques. Carilian-Goeury et V. Dalmont, Paris, 358p

Akitomo K, Tanaka K, Kumagai M, Jiao C (2009) Annual cycle of circulations in Lake Biwa, part 1: model validation. Limnology 10:105–118

Appleby PG, Flower RJ, Mackay AW, Rose NL (1998) Paleolimnological assessment of recent environmental change in Lake Baikal: sediment chronology. J Paleolimnol 20:119–133

Berger A (1978) Long-term variations of daily insolation and Quaternary climatic changes. J Atmos Sci 35(2):2362–2367

Berger A, Loutre MF (1991) Insolation values for the climate of the last 10 million years. Quat Sci Rev 10:297–317

Charlet F, Fagel N, De Batist M, Hauregard F, Minnebo B, Meischner D, Team SONIC (2005) Sedimentary dynamics on isolated highs in Lake Baikal: evidence from detailed high-resolution geophysical data and sediment cores. Glob Planet Change 46:125–144

Croll J (1875) Climate and time in their geological relations. D. Aplleton and Company, New York, 577p

Droxler AW, Poore RZ, Burckle LH (ed) (2003) Earth's climate and orbital eccentricity: the marine isotope stage 11 question. Geophys Geophysical Monograph Series 137, AGU, Washington, D. C., 240p

Dunne E, Lee LC, Reddington C, Carslaw KS (2012) No statistically significant effect of a short-term decrease in the nucleation rate on atmospheric aerosols. Atmos Chem Phys 12 (23):11573–11587

Eduard YO, Oleg MK (2010) Glaciers and meltwater flux to Lake Baikal during the Last Glacial Maximum. Palaeogeogr Palaeoclimatol Palaeoecol 294:4–15

Endoh S (1995) Review of geostrophic gyres. In: Okuda S, Imberger J, Kumagai M (eds) Physical processes in a large lake: Lake Biwa, Japan. AGU, Washington, 216p, pp 7–13

Endoh S, Okumura Y (1993) Gyre system in Lake Biwa derived from recent current measurements. Jpn J Limnol 54:191–197

Endoh S, Okumura Y, Okamoto I (1995) Field observation in the north basin. In: Okuda S, Imberger J, Kumagai M (eds) Physical processes in a large lake: Lake Biwa, Japan. AGU, Washington, 216p, pp 15–29

EGS (European Geophysical Society) (1995) Milutin Milanković 1879–1958. Katlenburg-Lindau, FRG, 181p

Hays J, Imbrie J, Shackleton N (1976) Variations in the Earth's orbit: Pacemaker of the ice ages. Science 194:1121–1132

Holmes A (1965) Principles of physical geology. Ronald Press Co., New York, 1303p (Revised Edition)

Horie S (1973) An outline of the paleolimnological works of Lake Biwa-ko, Japan. Jap J Limnol 34:49–54 (in Japanese with English abstract)

Horie S (ed) (1984) Lake Biwa. Dr. Junk Publ., Dordrecht, 654p

Horie S (ed) (1987) History of Lake Biwa. Kyoto University, Otsu, 242p

Imbrie J, Imbrie KP (1979) Ice Ages: solving the mystery. Harvard University Press, Cambridge, 224p

Imbrie J, Hays JD, Martinson DG, McIntyre A, Mix AC, Morley JJ, Pisias NG, Prell WL, Shackleton NJ (1984) The orbital theory of Pleistocene climate: Support from a revised chronology of the marine δ^{18}O record. In: Berger AL, Imbrie J, Hays J, Kukla G, Saltzman B (eds) Milankovitch and climate: understanding the response to astronomical forcing. Springer, New York, 896p, pp 269–305

Imbrie J, McIntyre LA, Mix AC. (1989) Oceanic response to orbital forcing in the late quaternary: observational and experimental strategies. In: Berger A, Schneider SH, Duplessy J-C (eds) Climate and geosciences, a challenge for science and society in the 21st Century. Kluwer Academic Publishers, Dordrecht, 724p

Ishikawa K (2004) Climato-limnological changes inferred from core sediments of Lake Biwa, Japan. Master's thesis for Kanazawa University, 62p (in Japanese with English abstract)

Karabanov E, Prokopenko A, Williams D, Khursevich G, Kuzmin M, Bezrukova E, Gvozdkov A (2003) High-resolution MIS 11 record from the continental sedimentary archive of Lake Baikal. In: Droxler AW, Poore RZ, Burckle LH (eds) Earth's climate and orbital eccentricity: the marine isotope stage 11 question. Geophysical Monograph Series 137 AGU, Washington, D. C., 240p, pp 223–230

Kashiwaya K (1994) A quantitative expression for external forces. In: Kirkby MJ (ed) Process models and theoretical geomorphology. Wiley, Chichester, 417p, pp 85–95

Kashiwaya K (2012) Earth surface processes and environmental changes in lake-catchment systems. Trans Jpn GeomorpholUnion 33:121–136

Kashiwaya K, Ochiai S, Sakai H, Kawai T (2003a) Onset of current Milankovitch-type climatic oscillations in Lake Baikal sediments at around 4 Ma. Earth Planet Sci Lett 213:185–190

Kashiwaya K, Ochiai S, Tsukahara H, Sakai H, Kawai T (2003b) Long-term late Cenozoic global environmental changes inferred from Lake Baikal sediments. In: Kashiwaya (ed) Long continental records from Lake Baikal. Springer, Tokyo, 370p, pp 1–20

Kashiwaya K, Ochiai S, Sumino G, Tsukamoto T, Szyniszewska A, Yamamoto M, Sakaguchi A, Hasebe N, Sakai H, Watanabe T, Kawai T (2010) Climato-hydrological fluctuations recorded in long lacustrine records in Lake Hövsgöl, Mongolia. Quat Int 219:178–187

Kashiwaya K, Yamamoto A, Fukuyama K (1991) Time variation of in coarse materials from lake bottom sediments and secular paleoclimatic change. Geophys Res Lett 18:1245–1248

Kirkby J (2008) Cosmic rays and climate. Surv Geophys 28:333–375

Krahenbuhl DS (2015) Investigating a solar influence on cloud cover using the North American regional reanalysis data. J Space Weather Space Clim 5:A11. doi:10.1051/swsc/2015012

Köppen W, Wegener A (1924) Die Klimate der Geologischen Vorzeit. Gebrüder Borntraeger, Berlin, 256p

Kumon F, Kamitani T, Sutoh K, Inouchi Y (1993) Grain size distribution of the surface sediments in Lake Biwa. Mem Geol Soc Jpn 39:53–60 (In Japanese with English abstract)

Laken BA, Pallé E, Čalogović J, Dunne EM (2012a) A cosmic ray-climate link and cloud observations. J Space Weather Space Clim 2:A18
Laken BA, Palle E, Miyahara H (2012b) A decade of the moderate resolution imaging spectroradiometer: is a solar cloud link detectable? J Clim 25(13):4430–4440
Laskar J, Joutel F, Boudin F (1993) Orbital, precessional, and insolation quantities for the Earth from -20 MYR to +10 MYR. Astron Astrophys 270:522–533
Laskar J, Robutel P, Joutel F, Gastineau M, Correia AC, Levrard B (2004) A long-term numerical solution for the insolation quantities of the Earth. Astron Astrophys 428:261–285
Laskar J, Fienga A, Gastineau M, Manche H (2011) La2010: a new orbital solution for the long-term motion of the Earth. Astron Astrophys 532:A89
Lockwood M (2012) Solar influence on global and regional climates. Surv Geophys 33:503–534
Milanković M (1920). Théorie mathématique des phénomènes thermiques produits par la radiation solaire. Académie Yougoslave des Sciences et des Arts de Zagreb, Gauthier-Villars, Paris, 339p
Milanković M (1941) Kanon der Erdbestrahlung und seine Andwendung auf das Eiszeitenproblem. Royal Serbian Academy, Belgrade, 633p
Ochiai S, Kashiwaya K (2003) Hydro-geomorphological changes and sedimentation processes printed in sediments from Lake Baikal. In: Kashiwaya K (ed) Long Continental Records from Lake Baikal. Springer, Tokyo, 370p, pp 297–312
Ochiai S, Kashiwaya K (2005) Climato-hydrological environment inferred from Lake Baikal sediments based on an automatic orbitally tuned age model. J Paleolimnol 33:303–311
Penck A, Brückner E (1909) Die Alpen im Eiszeitalter. Tauchnitz, Leipzig, 1199p
Pollard D (1983) A coupled climate-ice sheet model applied to the quaternary ice ages. J Geophys Res 88(C12):7705–7718
Polissar PJ, Abbott MB, Wolfe AP, Bezada M, Rull V, Bradley RS (2006) Solar modulation of little ice age climate in the tropical Andes. Proc Nat Acad Sci USA 103(24):8937–8942
Scheidegger AE (1970) Theoretical geomorphology, 2nd ed. Springer, Berlin, 435p
Shiki T, Kumon F, Inouchi Y, Kontani Y, Sakamoto T, Tateishi M, Matsubara H, Fukuyama K (2000) Sedimentary features of the seismo-turbidites, Lake Biwa, Japan. Sed Geol 135:37–50
Shimaraev MN, Verbolov VI, Granin N, Sherstayankin PP (1994) Physical limnology of Lake Baikal: a review. Baikal International Center for Ecological Research, Irkutsk/Okayama, 81p
Suda K, Seki K, Ishii J, Takatani S, Mizuuchi S (1926) The report of limnological observation in Lake Biwa-ko (I). Bull Kobe Marine Obs 8:1–103 (in Japanese)
Svensmark H (2007) Cosmoclimatology: a new theory emerges. Astron Geophys 48(1):1.18–1.24. doi:10.1111/j.1468-4004.2007.48118.x
Svensmark H (2015) Cosmic rays, clouds and climate. Europhys News 46(2):26–29. doi:10.1051/epn/2015204
Svensmark H, Calder N (2007) The chilling stars: a new theory of climate change. Icon Books, UK, US, Australia, 246p. www.iconbooks.co.uk
Svensmark J, Enghoff MB, Svensmark H (2012) Effects of cosmic ray decreases on cloud microphysics. Atmos Chem Phys Discuss 12:3595–3617
Svensmark H, Enghoff MB, Pedersen JOP (2013) Response of cloud condensation nuclei (>50 nm) to changes in ion-nucleation. Phys Lett A 377:2343–2347
Taishi H, Okuda S, Shiki T, Kashiwaya K (1991) A sedimentary anomaly and the related sedimentation process in Lake Biwa, Japan. Z Geomorph NF 83:241–249
Usoskin IG, Solanki SK, Kovaltsov GA (2007) Grand minima and maxima of solar activity: new observational constraints. Astron Astrophys 471:301–309
Usoskin IG, Korte M, Kovaltsov GA (2008) Role of centennial geomagnetic changes in local atmospheric ionization. Geophys Res Lett 35:L05811. doi:10.1029/2007GL033040
Vogel H, Meyer-Jacob C, Melles M, Brigham-Grette J, Andreev A, Wennrich VP, Tarasov PE, Rosén P (2013) Detailed insight into Arctic climatic variability during MIS 11c at Lake El'gygytgyn, NE Russia. Clim Past 9(4):1467–1479
Weertman J (1976) Milankovitch solar radiation variations and ice age ice sheet sizes. Nature 261:17–20

Yamamoto A (1995) Bottom sediments and paleo-hydrological processes. In: Okuda S, Imberger J, Kumagai M (eds) Physical processes in a large lake: Lake Biwa, Japan. AGU, Washington, 216p, pp 153–167

Yamamoto Y, Kashiwaya K, Fukuyama K (1984) Periodic variations of grain size in the 200 meter core samples from Lake Biwa. Trans Jpn Geomorphol Union 5:345–352 (in Japanese with English abstract)

Yamamoto A, Kashiwaya K, Fukuyama K (1985) Periodic variations of grain size in Pleistocene sediments in Lake Biwa and earth-orbital cycles. Geophys Res Lett 12:585–588

MIX
Papier aus verantwortungsvollen Quellen
Paper from responsible sources
FSC® C105338

If you have any concerns about our products,
you can contact us on
ProductSafety@springernature.com

In case Publisher is established outside the EU,
the EU authorized representative is:
Springer Nature Customer Service Center GmbH
Europaplatz 3, 69115 Heidelberg, Germany

Printed by Libri Plureos GmbH
in Hamburg, Germany